SAFETY ASSESSMENT OF EXISTING BUILDINGS AND STRUCTURES

Edited by

K.I.Eremin, V.D.Raizer, V.I Telichenko

ASV Construction
Stockholm, Sweden

Authors:
Adamenko I.A., Aleksandrov S.V., Alekseeva E.L., Babayan I.S., Banach A.V., Banach V.A., Bayburin A. KH., Belykh YU.V., Budko V.B., Eremin K.I., Grunin I.Y., Guriev A.A., Iriskulov A.R., Kalinin N.N., Kozachek V.G., Kulyabko V.V., Matveyushkin S.A., Merkulov A.F., Morozov P.A., Nikonov T.A., Pavlova G.A., Raizer V. D., Roitman V.M., Samolinov N.A., Senin N.I., Shuvalov A.N., Sushchev S.P., Sventikov A.A., Telichenko V.I.

Safety assessment of existing buildings and structures. Stockholm, Sweden: ASV Construction, 2016. – 268 p.

ISBN 978-91-982223-3-3

The largest structural failures of residential, public and industrial buildings stated the one of important humanitarian problem that is the safety protection of human life. The paper deals in particular with this problem.

Considering different approaches and methods the authors of this book submit the safety analysis of building structures in course of operation.

Authors are considering wide number of modern engineering problems, including survey, analysis and concept of structural failure, real bearing capacity of structures, methods and systems increasing structural safety.

This book is addressed to specialists involved in safety assessment of a wide range engineering structures in building industries, research organization and universities.

Preface

Assessing the safety of an existing structure may however differ very much that for a structure at the design stage. The existing buildings and structures are subject to deterioration and damage, they are likely to be periodically inspected and repaired or strengthened if necessary. Consequently, regarding the state of information the situation in assessing the existing buildings and structures is completely different from that during design. Moreover, special attention had to be paid to specific parts of the existing buildings, especially to those that are at high risk of damage according to observations of the building in service. On the other hand, the interpretation and analysis of additional information may not be simple. The assessment of actual safety of the existing buildings and structures will includes some actions discussed in this book.

This book contains the papers written by the specialists from research organizations, universities and regulatory agencies including Russia, Ukraine and Belarus. The authors were focused on applying the necessary in-situ data, experimental data and some necessary analytical rigor to the engineering context. In the book presented   new development as well as state-of –the –art applications of safety principles to all kind of buildings-residential, public, industrial. Moreover-in view of the growing concern for product liability – safety, reliability performance and quality assurance aspects of major structural schemes of buildings are covered.

There are two ways to evaluate the safety of existing buildings: the theoretical approach   and the empirical approach. The theoretical approach uses a rigorous mathematical theory generally known as reliability theory. However, the theory cannot be truly effective because the failure mechanism of most buildings is too complex from the theory to model, and the method requires supporting statistical data which often do not exist. The empirical approach, on the other hand, provides failure estimates based on empirical data. There are average estimates, applicable to large groups of near identical buildings. The empirical method cannot take into consideration specific features of a building which may distinguish the building from others of its class.

In this book, an approach to structural safety which may serve as a bridge between the theoretical and empirical approaches is exploded.

Results of the presented researches will definitely affect the future design practice.

K.I.Eremin,   V.D.Raizer, V.I. Telichenko.

3

Part I. Survey, Analysis and Concept of Structural Failure

Ch.1. Analysis of Causes and Consequences of ccidental Situations

Eremin K.I., Matveyuchkin S.A., Pavlov G.A., Alekseeva E.L., Senin N.I.[*]

Sec.1.1. General Comments

By "failure" we understand the local or full structural collapse that causes big financial and human losses. Investigation and analysis of happened failures testify that they take place usually as a result of identical reasons and mistakes. They can be:

- underestimation of the actual work of structure or it's fatigue;
- incompatibility or low quality of the materials used in the structures;
- poor quality or incomplete fulfillment of hydro-geological surveys, which in some cases leads to differential settlements of buildings;
- deviation from the established requirements and rules in the manufacture of construction and erection works, causing overstress and weakening of the structures and even the loss of their resistance.

Study, analysis and systematization of the surveys show that the failures in the majority of cases do not occur because of one reason but are the result in the aggregate number of reasons and the omission of specific features of structural work possessing different strength, durability and other physic-mechanical properties of structural materials. However, in each such case, the main and decisive cause of the collapse is often only one of the most pronounced miscalculations or defects in the design. In a modern industrial and year-round construction of the issues of accident-free and defect-free construction, increase of reliability and durability is one of the most urgent problems in the modern designing of buildings.

The size of the damage of structures and the degree of inflicted losses of the accident can be divided into major and minor (local) levels.

The largest include the accident and collapse, which cover the significant part of the building or its separate parts, which resulted in the cease work not only on this site, but also on adjacent to it. Limited destructions

---

[*] OOO "WELD" Co. Magnitogorsk, Chelyabinsk region, Moscow State University for Civil Engineering, Moscow

is caused by events local in nature, which do not violate the work of other related structures.

Modern check the status of structures using non-destructive methods of control and systematic monitoring of the buildings and facilities allow preventing the occurrence of accident.

Sec.1.2. Classification of Accidents

Several million tons of steel structures are operating in Russia at the present time. Although the quality if the structures and facilities is high, , there might occur issues during operation that might lead to accidents. To solve this problem, a comprehensive study of failures and damages of metal structures is necessary. In the works of different authors there are found attempts to classify accidents and their causes. So, the first attempt to classify was held in 1953by F.D.Dmitriev [1.1]. Dmitriev has allocated 3 causes of accidents:

- natural forces;
- imperfection of the engineering-technical knowledge;
- socio-economic conditions.

In [1.2] he listed: ignorance; savings; negligence; natural disaster.

In [1.3] he stated: the effects of force; mechanical or physical loads; chemical actions.

In [1.4] he discussed error design; defects arising in the process of execution of work; defects of the operation; the lack of study of working conditions and the properties of materials.

In [1.5] he considered: overloading as a result of underestimation of the actual load; the loss of stability (general, local, bending twist); unsuccessful design solutions and deviations from the project; the irregularities committed during manufacturing and installation of structures; the irregularities committed in the operation of the structures; accident as a result of metal fatigue and vibration; defectiveness of grounds; unforeseen reasons;

In work by K.I. Eremin and S.A. Nishcheta they noted the mistakes of the design; defects in manufacture, transportation and installation; the damage received by structures in irresponsible exploitation; accident conditions of the irresponsible use of the equipment; natural disasters.

As the causes of accidents are diverse and varied, moreover, there are no two identical accidents; there should be different classifications based on different parameters.

The authors proposed the following approach to the classification of accidents:

1. By types of failures:

- emergency destruction, characterized by sharp changes of state of the structure;
- gradual failure.

2. On subjectivity reasons:
- errors in the design of the structure;
- mistakes during installation design;
- errors in the operation.

3. By objective reasons.
- natural disasters;
- accident workshop equipment;
- structural wear in normal operation.

4. According to the type of limit state:
- ultimate limit state;
- serviceability limit state.

5. The factors that led to the occurrence of the limit state:
- the factors that contributed to the reduction of the limit values of the parameter: chemical actions (including corrosion); cracks;
- change of physic-chemical properties of a material as a result of aging;
- reduction of geometric characteristics of the cross-section of the element (cutouts, deflections);
- the factors that contributed to the increase in the values of designed parameters: increase of the own weight of the design (most often due to design errors); increase of operating loads; change of the geometrical scheme design.

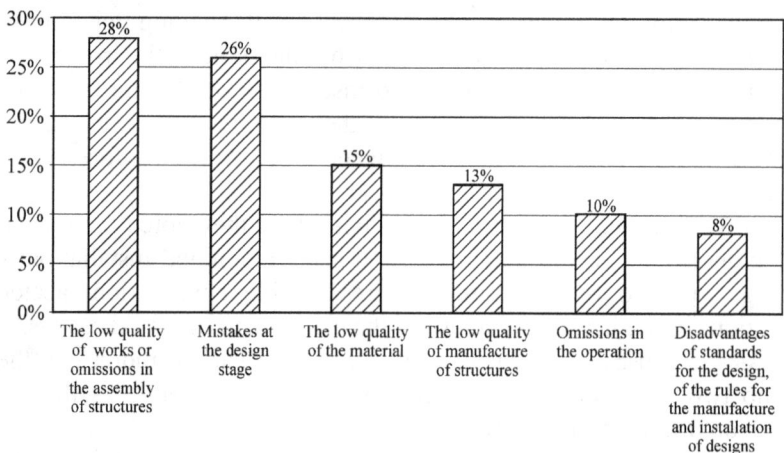

Fig.1.1. The distribution of the accidents in industrial buildings with steel framework

The present classification of the causes of accidents will give a more profound analysis of the causes and consequences of accidents, avoid the recurrence of similar accidents at various sites, and thus avoid suspension of production, the economic losses and, most importantly, human casualties.

According to the results of surveys , there has been established a percentage of the causes of accidental collapses of metal frames of industrial buildings in metallurgical complexes (Fig.1.1). In almost all of these cases the risk of occurrence of an emergency situation was not assessed in case of changed operating conditions and any damage in the bearing structures.

Sec.1.3. Accidents Associated with Collapse of Covering

The peculiarity of accidental structural collapse lies in the fact that the destruction occurs instantly. The scenario of events which led to the collapse may be different, as evidenced by the examples of some of the accidents that have occurred at the enterprises of the Chelyabinsk region for the past decade.

2001. The destruction of the part of the covering of warehouse building in OJSC "Zlatoust metallurgical plant". There were no victims. The reason of destruction was the increasing loadion the building coating.

2001. The destruction of the disk coating of the building of OJSC "Zlatoust metallurgical plant". There were victims. The main cause of the destruction was the low quality of steel load-bearing structures.

2006. The destruction of the structures of coverage within a temperature block of the building for rolling shop №5 of JSC "Magnitogorsk metallurgical combine". The area of the collapse is equal to 5000 m². There were victims. Before the destruction there was a fire, as a result of high-temperature impact on one of the trusses, after that the collapse of the adjacent trusses within the temperature block happened.

2006. The destruction of the unit covering the building of rotating furnaces OJSC "Magnitogorsk cement works". The area of the collapse was about 1760 m². There were no victims. The cause of the disaster was the increase of the loading on the roof as the result of the cement dust.

2006. The destruction of 80% of building structures of rotating furnaces OJSC "Korkinskiy cement plant". There were victims.

The most complicated task in the part of ensuring the structural safety is to predict their status and define the moment of termination in operation. The destruction is irrespective from the technical condition of structures. In this case, accident action which leads to collapse is a rare event, and the structure was not designed to this action.

Sec.1.4. Reliability Analysis of Coatings with Bearing Steel Structures

Qualitative analysis of structural reliability carried out on the basis of the method of structural schemes of reliability (GOST 51901.14-2005) and of analysis of fault tree method (GOST 51901.13-2005).

These methods were used for the identification the ways of realization a hazardous event and determination the logical connection between components, which ensure the system carrying capacity.

The analysis shall apply the following terminology:
- the system of elements - bearing and bracing elements of coating framework without expansion joints or its part within a temperature block;
- elements of the system - the trusses and bracing elements of the coating (including girders);
- failure of the element is a breakdown of the efficiency of the element associated with its destruction;
- failure of the system - the total destruction of the building coating.

Under system failure the collapse of the entire cover is taken, that is, the collapse of one or more trusses is the failure of individual elements of a system, not the system as a whole.

Failure of the system's elements or the system as a whole are considered to be the destruction of structure, after which it can't be restored.

At this stage of the analysis the causes that lead to failures of individual elements of the system are not addressing. The analysis is carried out by determining the sequence of failures of individual elements that lead to the top of the event, namely, a system failure.

Procedure for completiong of the tree fault was carried out with the application ""direct cause"" concept. Fig.1.2 presents the tree fault's system ""the cover of framework "". The analysis of the scheme allows to make the following conclusions:
- the events that are associated with the simultaneous failure of all columns or trusses, are unlikely;
- the most probable is a chain of events, which will start with the failure of one of the elements - a truss or column;
- the necessary condition for the collapse of the coating is not an indestructibility of the elements of all braces at a time but the continuity of the collapse of all trusses "one by one" (the progressive collapse of structures).

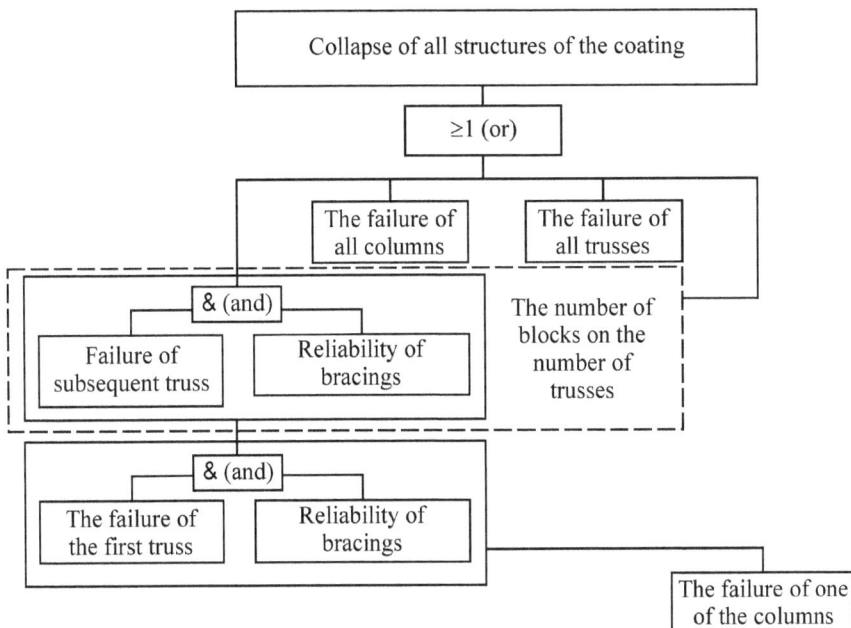

Fig.1.2. Tree fault system

The system represents the following model: the basic elements of the system are the trusses, and the bracing elements define the scheme of interaction of the individual elements among themselves. The first model represents in series diagram which ensured reliability of connecting elements in time of the failure of the trusses. The second model is a parallel diagram, that is, the failure of the trussed rafters leads to failure of the adjoining connecting elements, but simultaneously the adjacent trussed rafters survive. A third model - parallel-in series diagram, in this case the failure of one trussed rafters leads to failure of the trusses within one of the blocks.

For systems, consisting from n elements, considering that these elements have the same probability of no failure (PNF) for each of them, we get the following values of PNF to the system of "the covering of framework" depending on the type of structure:

– for in serial $R_s = R^n$;

– for parallel: $R_s = 1 - (1 - R)^n$; 2

– for parallel-in serial: $R_s = 1 - (1 - R^n)^x$.

Let us state the reliability of a single rafter truss on the basis of the following conditions –reliability index β is changing from 1.65 to 3,11 [1.9], the PNF will take the value in the interval from 0,950 to 0.999.

9

The most applicable unified step of size for truss coatings of industrial buildings is of 6.0 m. Thus, you can set the maximum number of trusses within a temperature block from 20 to 40 units.

The reliability of "in series" system will be determined by the number of elements and by the reliability of individual one; which means thatthere are two ways to increase the reliability of the system (Fig.1.3):

- reduce the number of elements;
- increase the reliability of the individual elements.

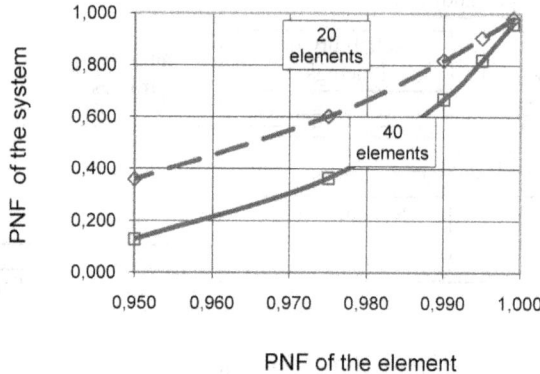

PNF of the element

Fig.1.3. Change of the reliability in series system

Consider the reliability of parallel-in series system in conditions of "in series" scheme, by splitting into blocks with the same number of elements. Fig.1.4 shows the change of system reliability, consisting of 20 elements, with different initial reliability of the individual elements.

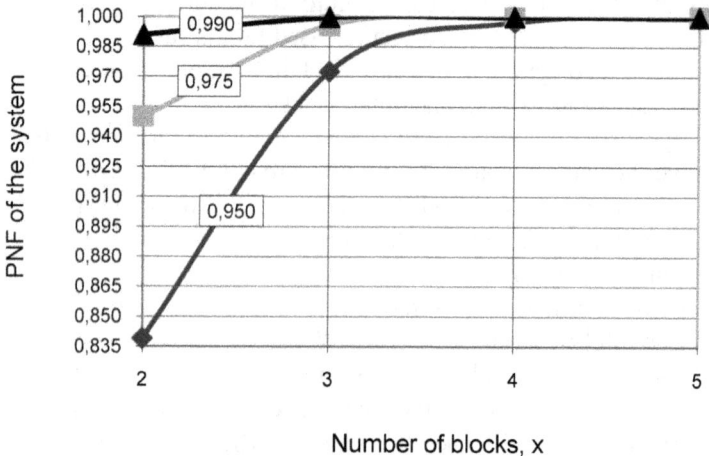

Number of blocks, x

Fig.1.4. Change of reliability of parallel-sequential system

On the basis of the received data we can draw the following conclusion - in systems consisting of four blocks, connected in parallel, the effect of initial reliability of individual elements on the reliability of the system as a whole is minimized. The parallel block diagram is ideal in relation to the others, but it is an assumption that the consequences in case of collapse of one truss always less than the collapse of the whole.

Sec.1.5. Risk Analysis of Accidents in Bearing Coatings

The main difficulty in assessing the reliability of real structures with the use of suggested methods is the definition of allowable and actual indexes of reliability. In addition, adopted to be equal for all elements of the system the probability of no failure (PNF) will be, undoubtedly, different in real structures on the ground of defects and damages accumulated in time of the operation.

According to [1.10] the risk is the combination of the probability of an event and its consequences. Therefore at the following stage of the analysis of these systems the issues of risk are taken into consideration, as in this case, apart from the probabilistic assessments, there exist losses and their estimate is more important.

Considered event is the probability of failure of the system [1.11], $F_S = 1 - R_S$ and the numerical characteristics of which can be determined by the PNF.

In the general case, the concept of losses can be considered as the failure associated with the physical damage to human health, property or the environment [1.10]. As the losses U at this risk level of accidental coating collapse, the relative value describing the area of the collapse is considered. It should be added that a more detailed risk analysis must take into account all the components of the failure associated with the infliction harm to human health, the technology and the environment, but in this research we will limit ourselves to the adopted relative value. In this case, for the considered system when collapse includes all trusses the relative size of the collapse will be equal to1. Such loss is typical for in series systems, then numerically the risk $-r$ is equal to the probability of failure F_s. For parallel-serial system consisting of two equal size blocks, the loss will be 0.5 from the maximum possible area of the collapse. Thus, to make a multiple reducing the number of blocks of the system, the graph of the relative square of the collapse was completed (Fig.1.5).

The value of risk considers as the product of the probability of failure to the losses: $r = F_S \cdot U$ .Change of these parameters is presented in Fig.1. 6.

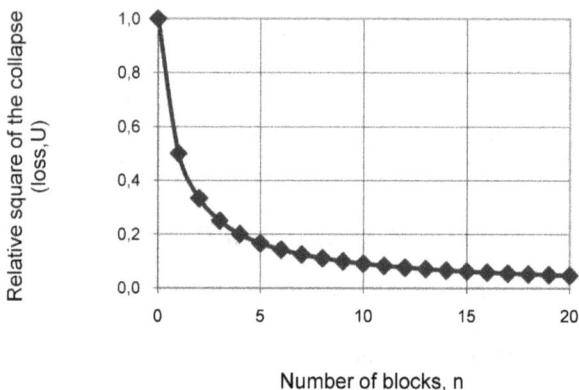

Fig.1.5. Variation of the amount of losses

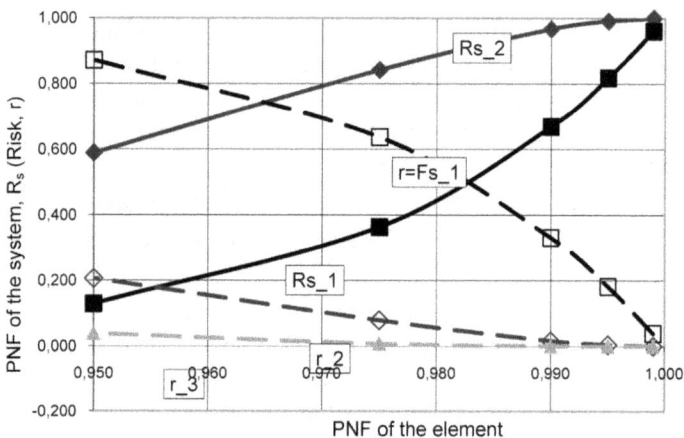

Fig.1.6. Assessment of reliability and risk: *1* – connection in series; *2*– parallel-serial, consisting of two blocks; *3* – parallel-serial, consisting of three blocks

Analyzing the obtained results we can note the following feature: in order to minimize risks in series system, it is necessary to increase the reliability of individual elements to the values close to unity, or change the structure of the system to the parallel-in series, without changing the reliability of the individual elements.

It was stated that on the basis of the results on reliability and risk analysis for accidental collapse of the bearing metal structures of coatings the minimization of failures will be possible by means of regulation of interaction among the load-bearing elements of coatings. There is no need to change the established approaches to ensure the reliability of bearing structures of coatings, as in this case, the trusses are highly reliable.

It is enough only to implement functional arrangements with respect to bracing elements of the coatings, which should be as follows. It is obligatory to consider two design situations [1.8]:

- persistent – with duration of the same order of as the structure's service life;
- accidental – with short durations and low occurrence of probability but very important from the point of view of the consequences of achievement of limit states.

Accidental situation, in this case, connects with the collapse of one of the trusses.

The design of bracings in an accidental situation must work in such a way as to prevent the collapse of the adjacent truss.

Sec.1.6. Ensuring of Safe Operation in Civil Buildings

The analysis of the reasons of accidental collapse for constructed and operated buildings was installed on the base of [1.15, 1.16]:

– proactive measures for the prevention of accidents, providing the safety to the constructed and operated buildings taken by the bodies of executive power and bodies of supervision, as well as the construction and operational companies, enterprises and associations, are not sufficient;

– the number of accidents is not decreasing, their seriousness increases and the number of victims of accidents increases as well;

– the majority of accidents belong to the period of operation and makes up 85% of the total number of registered accidents. The materials of investigation show that the main causes of accidents on the operated facilities are gross violations of the rules of technical operation of buildings;

– there is no proper control over the technical condition of buildings;

– the accident at the operated facilities is not only caused by the unsatisfactory operation of buildings. Analysis of the causes of accidents shows that almost always the main or associated cause was in violations on the stage of design or construction.

In Fig.1.7 there is the number of registered accidents in the territory of the Russian Federation [1.15, 1.16].

Analysis of the reasons of accidental destruction of the building structure, carried out on the basis of official data [1.15, 1.16] and also based on the results of the investigation received by the specialists of "WELD" Ltd. and allows to identify the following main reasons:

- violation of the rules of exploitation;
- defects at the stage of construction and deviations from the projects;
- violations of the technology in construction, reconstruction and repair;
- low quality of structural manufacturing.

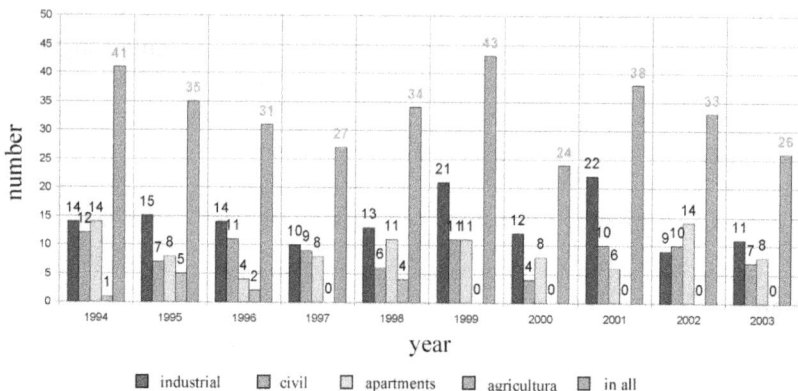

Fig.1.7. The number of registered accidents, occurred in the territory of the Russian Federation

Table. 1.1

Presents the results of the distribution of accidental causes

The source	The causes of accidental destruction, %		
	Violation of the rules of operation	Defects at the stage of construction and deviations from the projects	Other
The official data 1993 - 1998 [17]	35	26	39
The official data 1998 - 2003 [18]	35	35	30
Data "WELD" LTD. Co 1993 – 2010	26	28	46

The dynamics of changes of accidental destruction reasons according to the official data [15, 16] for the period from 1981 to 2003 is shown in Fig.1.8. Analysis of the data on the dynamics of accidental destruction variation shows that the proportion of the accidents, connected with violation of the rules of operation, is constantly growing (Fig.1.9).

14

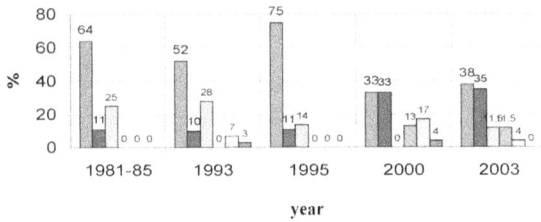

Fig.1.8. Dynamics of changes of the accidents causes.
(In percentage to total quantity of accidents)

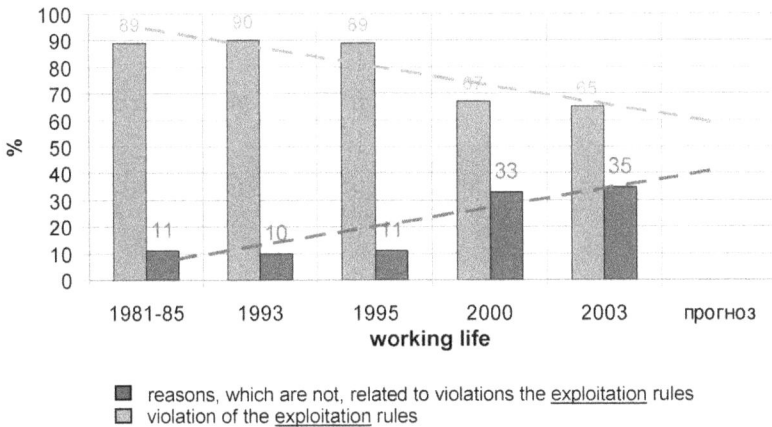

■ reasons, which are not, related to violations the exploitation rules
☐ violation of the exploitation rules

Fig.1.9. Analysis of the data according to the dynamics change of the causes
of accidents

The number of graphs is presented based on official data. They characterize changes in the probability of the number of accidents per year according to the total number of major accidents, including also accidents with human victims. Statistical analysis of accident data for different periods of time showed that the probability of consequences of the accident with fatalities was approximately equal to: $\approx 0, 3$ (Fig.1.10).

15

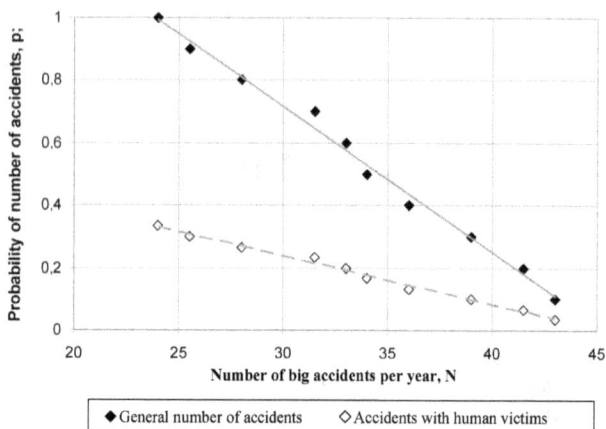

Fig.1.10. Probability of the accidents number

The period from 1993 to 2010 is characterized in a steady increase in the number of accidents, due to violation of the rules of operation or illiterate exploitation of buildings. At this to 30% of the accidents with structures are associated with human victims.

Sec.1.7. Overview of Accidents in Civil Buildings

Overview of accidents has been performed on the example of the buildings intended for the institutions of education, analyzed on the basis of published materials [1.15, 1.16] and Internet information sources [1.20]. A total of 34 examples of accidental destruction in the period from 1998 to 2010, of which 22 case occurred in Russia, 3 in Ukraine, 2 in Kazakhstan and 7 in China and Australia were considered. Some examples are presented in the photos (Fig.1.11, 1.12).

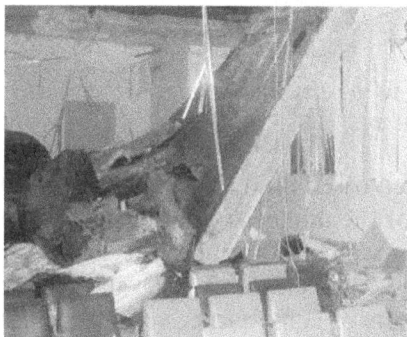

Fig.1.11. The collapse of the floor in the school building in Ryazanovka settlement, Kazakhstan

16

Fig.1.12. The collapse of the building's roof, floor and part of the walls of the façade in the building of the Lyceum in Kirovskoe town, Donetsk region, Ukraine

Generalization of the received data allows to distinguish the following characteristic features of the occurred accidents:

– up to 82% of the accidents resulted from the collapse of the floor structures, from them only 26% was accompanied by the collapse of the walls;

– up to 9% of the accidents resulted from the collapse of the walls;

– up to 9% of the accidents resulted from the collapse of the structures of the roofs, half of which is not connected with the collapse of the low arranged (walls, floors) structures.

The main causes of failure are:

– violation of the rules of operation (non-fulfillment of proper control over the condition of structures and engineering systems, the absence or improper performance of current repairs);

– violation of the technology of construction in reconstruction or repair;

– unaccounted or changed during the process of operation snow loads (as a rule) and other climatic actions;

– deviations from the design during the construction of buildings.

It should be noted that to the accidental destruction of structures leads, as a rule, a complex of reasons, the main of which are listed above. Diagrams of data analysis on accidental destruction in buildings are presented at Fig.1.13 and 1.14.

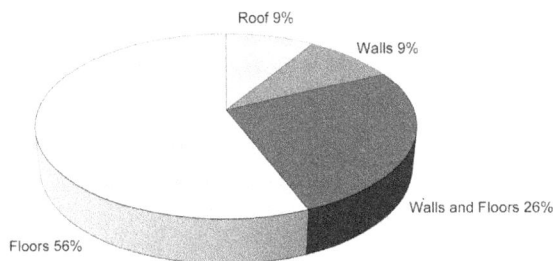

Fig.1.13. Distribution of collapses on the types of structural elements

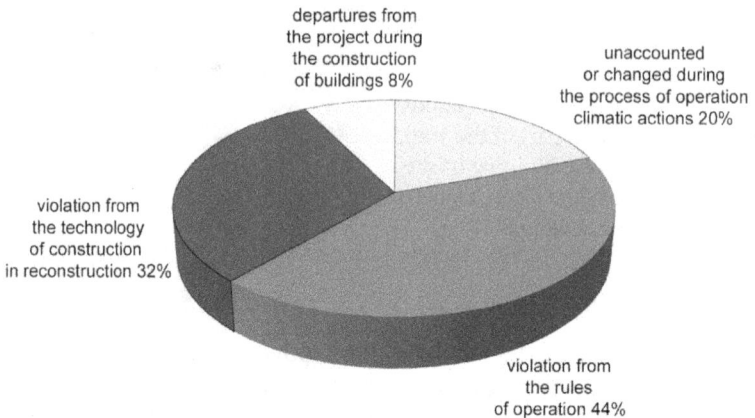

departures from
the project during
the construction
of buildings 8%

unaccounted
or changed during
the process of operation
climatic actions 20%

violation from
the technology
of construction
in reconstruction 32%

violation from
the rules
of operation 44%

Fig.1.14. Distribution of the main causes of collapse

It should be noted that the data placed in the mass media, does not always contain all information, necessary for analysis. This data often is not the result of the work of the specialists investigating the causes of accidents. However, in the absence of other information it is necessary to study and summarize all available information for the solution of the problem.

Sec.1.8. Damage of Bearing Structures in Civil Buildings

[1.21] describes safety issues of strategically important buildings, which are classified as the largest production building area of more than 2×10^4 m^2 with crane equipment and with carrying capacity of more than 500 tons in shipbuilding, aviation, rocket and space complexes, offshore installations, unique bridges and tunnels longer than 3000 m on strategic highways, covered buildings of various purposes with large masses of visitors (more than 2×10^4 people) and other structures. There is also information on the frequency of occurrence and losses from accidents and catastrophes on the objects of various purposes depending on their serial production, "from large-scale to unique".

The buildings of the educational institutions do not fall into the category of "premises with large masses of the people more than 2×10^4 people". The mass character of the visitors of the building of the educational institutions corresponds to the category of "not more than 10^3 people". However, accidents at a similar object, associated with fatalities, have a serious social impact, and especially if it happens on sites related to mass and long-term stay of people, most of them are children and young people. And, unfortunately, such examples are known.

18

To solve the assessment problem of the damage in the buildings of the educational institutions it is necessary to summarize survey results of the objects, conducted in 2005 - 2010. Studied objects were located in Russia-Moscow, Chelyabinsk, Sverdlovsk, Samara, Rostov, Perm, Primorskyi, Tomsk regions and in Bashkortostan Republic. In total, 30 buildings were examined.

It was found that according to the results of the survey almost all of the buildings have defects and damage of structural elements. Number of buildings with defects of structural elements is shown at Fig.1.15.

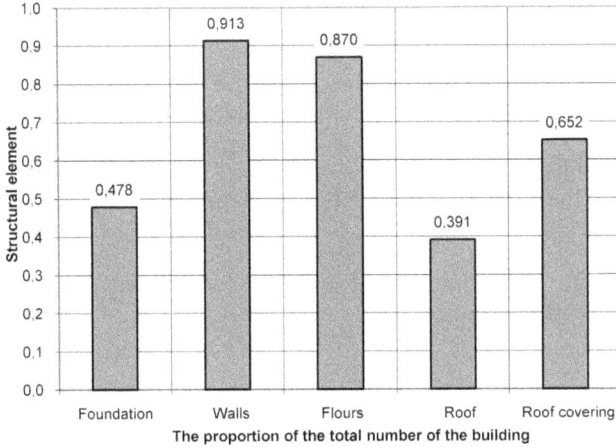

Fig.1.15. Buildings with defects

There has been carried out an analysis of damaged structures for building of educational institutions. Defects and damages were grouped on the following grounds:
– by structural accessories;
– by nature of a defect or damage.

Structural accessories the following as the main load-bearing elements of the surveyed buildings were identified:
– foundations;
– load-bearing walls;
– floors (basements, intermediates, attics);
– load-bearing structures of the roofs.

By the look of defects and damage the following groups were defined:
– cracks in structures;
– crazing (including denudation of reinforcement);
– destruction of surface (weathering of the solution, loss of individual bricks, etc.);

19

– damage of the ornamental or decorative coatings;

– soaking of structures (atmospheric moisture, groundwater, when leakage from engineering networks of buildings);

– corrosion of reinforcement.

As a result obtained from the survey analysis there was created a polygon (Fig.1.16) for the relative frequency of occurrence of defects . As the relative frequency the accepted value was admitted with correspondence to the ratio of the number of this type of defect at all surveyed facilities to the total number of detected defects in all objects:

$$w_j = \frac{\sum m_i}{n_{\sum}} \tag{1.1}$$

Where w_j is the relative frequency of occurrence of the defect or damage to the j-the type;

$\sum m_i$ is the number of defects and damage of this kind on all objects;

$n_{\sum}$ – the total number of defects and damage at all facilities; i - the number of the object.

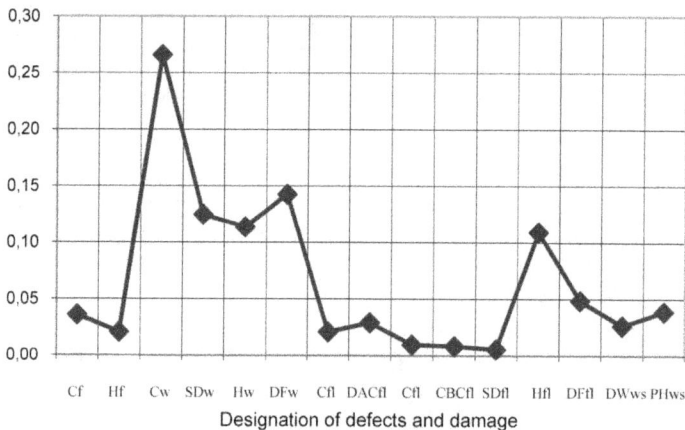

Fig.1.16. Polygon for the relative frequency of occurrence of defects

In Fig.1.16 the following designations of defects and damages are accepted: the C – crack; H – soaking; SD – surface destruction; CR – crazing; DAC – denudation of reinforcement with corrosion of armature; CBC – cracks in concrete with corrosion of armature; DF - destruction of the finishing coatings; DW – digestion; PH – destruction of the elements of furring, flooring; f – foundations; w - walls; fl – floors; ws – wooden structures. For example – *Cf* – crack in foundation, *SDfl* - floor surface destruction.

Frequency of occurrence of damage was determined based on the obtained data, taking into account duration of operation of the investigated objects.. Information in Table.1.2 provides the distribution of the surveyed buildings, including the average of the frequency of occurrence of damage in the bearing structures of the buildings of the educational institutions.

Table 1.2

Information about damage of structures

№	Indicator	Units of measurement	Value
1	The maximum duration of operation at the time of the survey	Years	100
2	The maximum duration of operation at the time of the survey		14
3	The average duration of operation		47
4	The average number of injuries (*d*) to the same object, including:	Piece	31
	– foundations		2
	– walls		20
	– flours		7
	– other structures		2
5	The average frequency (f) the occurrence of damage to one of the objects, including:	pieces/year	0,66
	– foundations		0,04
	– walls		0,43
	– flours		0,15
	– other structures		0,04

Considering that the damage in the majority of structures is identified visually, it's possible to set minimum frequency of inspection of structures for educational institute based on the obtained statistical data.

$$t_c = \frac{1}{f_c}, \qquad (1.2)$$

Where t_c – is the frequency of inspection for structures; f_c –is the frequency of occurrence of damage depending on the type of structures.

The minimum frequency of surveys of buildings of educational institutions is presented in Table 1. 3.

Table 1.3

The frequency of inspection of structures

№	Type of structural element	The frequency of inspection, years
1	Total frequency for the whole building	1,5
2	Foundations	25,0
3	Walls	2,3
4	Fours	6,7
5	Other structures	25,0

On the basis of the obtained data diagram (fig.1.17) shows assessments of the quality of building operation with indicator "the minimum frequency of inspections".

Permissible limit

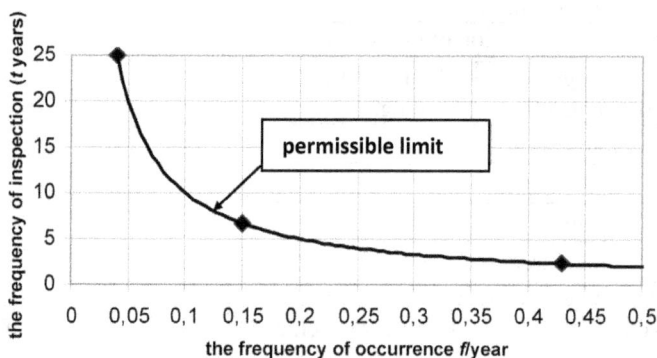

Fig.1.17. The dependence of the minimum frequency of examinations from the frequency of occurrence of damage

Most of the civil buildings performed without the use of technically complex or unique designs, such objects are not experiencing extreme influences in the process of exploitation, which are characteristic of the objects of industrial purpose. Accumulation and development of damage (up to 80%) on such objects, as a rule, occurs because of an improper or incorrect operation. Timely and professionally conducted inspection is a serious condition for ensuring safe operation of civil buildings.

Sec.1.9. Analysis of Accidents at Metallurgy Factories

At the metallurgical complexes you can currently find a significant physical deterioration of the sheet linearly extended steel structures, low

22

level of technological security, which inevitably leads to the occurring of structural destruction. There are some examples.

In recent years one of the most notable accidents occurred at the Ukrainian enterprise "Azovstal" (Mariupol) on March 23, 2006. The accident took place when the blast furnace №3 was ready to stop for a planned overhaul. The explosion of the gas mixture in the dome scrubber furnace has lead to the collapse of massive steel structures.

On February 9, 2007 there was an explosion of gas on the main gas control unit in oxy-gas shop of JSC "Amur-Steel", where as a result a door of the station management was rejected on 18 m, punching a hole in a metal fence of the floor of the gas regulation points. The shock wave of the explosion destroyed capital separation wall, and the ruins of it's debris ended up in the regulatory room.

On March 11, 2010 in the coke shop of LLC "Chelyabinsk plant for the production of coke and chemical products" (LLC "Mechel-Koks") an explosion occurred with the subsequent fire in the course of the planned steaming gas pipeline of the coke batteries. The explosion and subsequent fire destroyed the lower compartment of the coke-oven battery and damaged the production spans length about 60 m, with one of the workers in the shop was killed and two others were injured.

In recent years the number of man-made emergency situations (ES) still remains at a high level (Fig.1.18, 1.19). The most serious consequences of damage and the amount of victims occur in the fire hazardous industries, such as metallurgical industry.

The number of incidents in the metallurgical production remains at a constantly high level, as evidenced by the data from the Gas Rescue S(GRS) of the metallurgical enterprises (Fig.1.20).

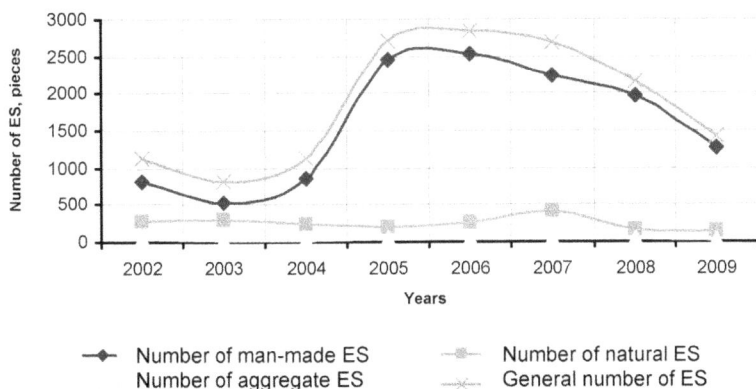

Fig.1.18. Distribution of the number of emergency situations in years

23

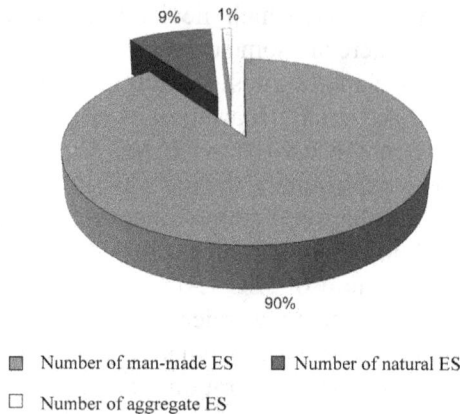

9% 1%

90%

■ Number of man-made ES   ■ Number of natural ES
□ Number of aggregate ES

Fig.1.19. Distribution of the number of emergency situations by sight

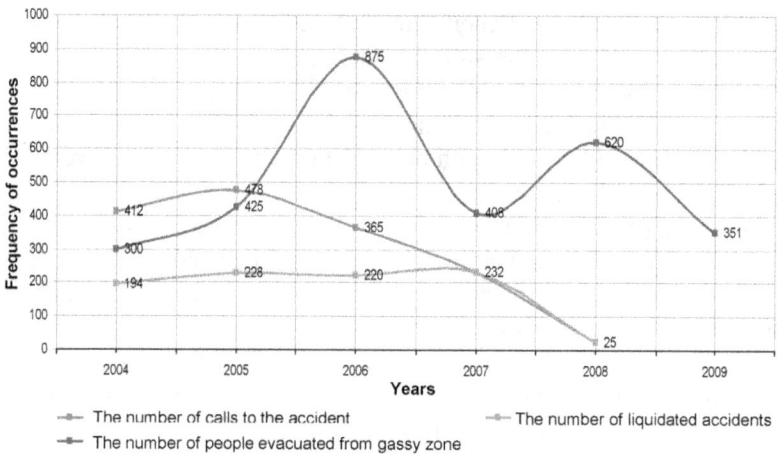

—■— The number of calls to the accident   —■— The number of liquidated accidents
—■— The number of people evacuated from gassy zone

Fig.1.20. Number of accidents

The analysis of fatal accidents has shown that their main reasons were:

- unsatisfactory organization and factory operation (60%);
- failure of the equipment (30%) or lack of technological instructions in the management of metallurgical processes (10%) (Fig.1.21).

The state of accident rate and industrial safety at metallurgical and coke-chemical enterprises adversely has been affected by the following factors:

– physical depreciation of technological equipment;

– ill-timed and sub-quality of capital and current repairs of equipment, buildings and structures;

– operation of the equipment with spent regulatory period;
– use of imperfect technologies;
– uncontrolled reduction of the number of skilled technicians and production personnel;
– reduction of the quality of vocational training of the staff. Also, the analysis of accidents and injuries showed that the main causes of accidents with the sheet linearly extended metal structures were disadvantages and violations during the construction and operation of the equipment.

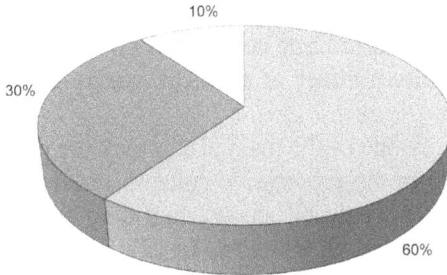

Fig.1.21. The main causes of injury

☐ unsatisfactory organization and realization of works
☐ failure of the equipment
☐ lack of technological instructions

Analysis of the causes of accidents confirms that they occur in the unfavorable combination of several factors. At the same time in all cases the source of destruction was damage accumulated in the process of exploitation.

The analysis of results of technical diagnostics, for assessment of the defectiveness of the exploited sheet linearly extended metal structures was made by specialists of the company "WELD" and the Moscow State University of Civil Engineering (MGSU) in the period from 2000 to 2010 inclusive. The main defects of the operation are cracks that occur at the same time with the general and local corrosions.

Analysis of the causes of accidents and accumulated flaw with exploited metal structures has shown the need for adjustments of existing approaches to assessing the safety of these facilities. Ensuring reliability and safety of potentially dangerous objects is at the present time extremely sharp and topical issue. Two directions are mainly used now:
– constant monitoring of the technical condition;
– establishment of a technical condition of structures on the basis of application of non-destructive control modern methods with the assessment of a residual resource and determination of the period of its further safe operation.

Ch.2. Resistance of Buildings in Case of Accidental Actions Including Fire

Roitman V.M., Telichenko V.I.[*]

Sec.2.1. General Comments

The problem of ensuring the safety of high-rise buildings and other unique objects in the light of the terrorist threat is very urgent, since these objects are among the most vulnerable.

The tragic events of September 11, 2001, associated with the terrorist attack at high-rise buildings of the world trade center (WTC) and the Pentagon building, set before humanity number of political, social and technical problems [2.1-2.5].

Among the technical challenges one of the main places took the problem of protection of unique objects from the progressive collapse in emergency situations, connected with complex of accidental actions, including fire.

"The combination of special effects" is invited to call the combination of operating loads and "force majeure" additional loads on the structures of objects in times of emergency.

In the papers [2.2-2.4] definitions have been formulated for the following concepts:

An accidental action on the building - exclusive effect, sharply different from the regular conditions of existence. Basic man-made accidental actions on buildings are: blow (I), the explosion (E), fire (F), the load (S), etc. The combined accidental actions - emergency situation connected with the origin and development of several types of special effects to an object in different combinations and sequences. As an abbreviation of this concept the name is used "combined hazardous effect", (CHE) [2.2-2.5]. Combined accidental actions with the participation of fire - emergency situations, connected with the emergence and development of several types of special effects to an object in different combinations and sequences, and one of those actions is fire load.

Sec.2.2. The Fire on Ostankino TV Tower on August, 2000

Ostankino TV tower was built in 1967 and was designed by Great Russian scientist and engineer N.V. Nikitin. The height of the tower is 533 m. The main tower's structure is reinforced concrete hollow conical shell with a strong base. The upper part of the tower (with a mark 385 m) is an antenna in the form of a steel telescopic pipe 148 m high (Fig.2.1).

---

[*] Moscow State University of Civil Engineering

Thickness of walls of the tower is 350-400 mm. The thickness of the concrete protective layer in the main reinforcement of the tower stem was 60 mm.In the internal space of a reinforced concrete shell of the stem (Fig.2.2) throughout the height placed engineering communications and vertical transport: four elevators shafts, shaft power of communication cables, radio feeders, systems of water supply and sanitation [2.8, 2.9].

The shell of the tower on the whole height was tightening by pre-strained cable fittings, which connects to the special level near the inner surface of the stem. These tensile rope elements were designed to be open and were inside of the tower at a distance of 2-5 cm from the surface of the inner wall of the trunk.

Fig.2.1. General view of the Ostankino TV tower

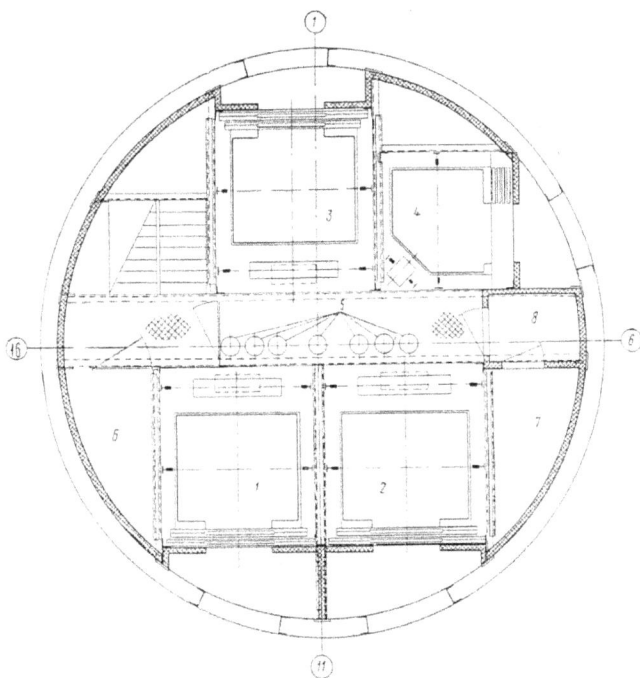

Fig.2.2. Plan of the trunk of Ostankino TV tower: *1–4* – elevators; *5* – radio-feeders; *6* – mine the cable connection; *7* - mine cables; *8* – mine plumbing

27

Fig.2.3. The fire on the Ostankino TV
tower

The fire on the Ostankino television tower took place in the night of August 28th, 2000. The fire started in the area of marks +454 – +430 m of the antenna part of the tower (Fig.2.3). The spread of fire took place from the top down to the level of +80 m (Fig.2.4). Values of the temperature of the fire and the time of its action on the height of the fire can be found in Fig.2.4.

During the fire on the Ostankino TV tower firefighters and engineers had to decide will it the tower collapse right away or will it be a progressive collapse during a fire or directly after it or if there is no such danger?

The answer to this question depended on the whole complex of very important decisions, including the urgent evacuation of people from the zone of the possible fall of the wreckage of the tower (Fig.2.5).

Experts divided on this issue , but in the end, came to the conclusion that the tower would stay and the progressive destruction should not happen.

It was tough decision proven to be correct - the progressive collapse of the tower did not happen. It hasn't ha happened because of the correct tower fire resistance.design.

The value of the actual limit of fire resistance of the tower for the loss of the bearing capacity was more than 180 minutes. This limit of fire resistance of the tower with a reserve provided the necessary resistance of the tower during real fire on 27 August 2000, and the temperature of it was more "soft" compared to the one of the "regular" fire. Fire resistance of the tower provided the structural integrity of the building in time of fire. The shape of the shell of the tower allowed to continue operation of the tower after the recovery of the majority of pre-stressed wire ropes inside the trunk, which in the event of fire lost its functional purpose.

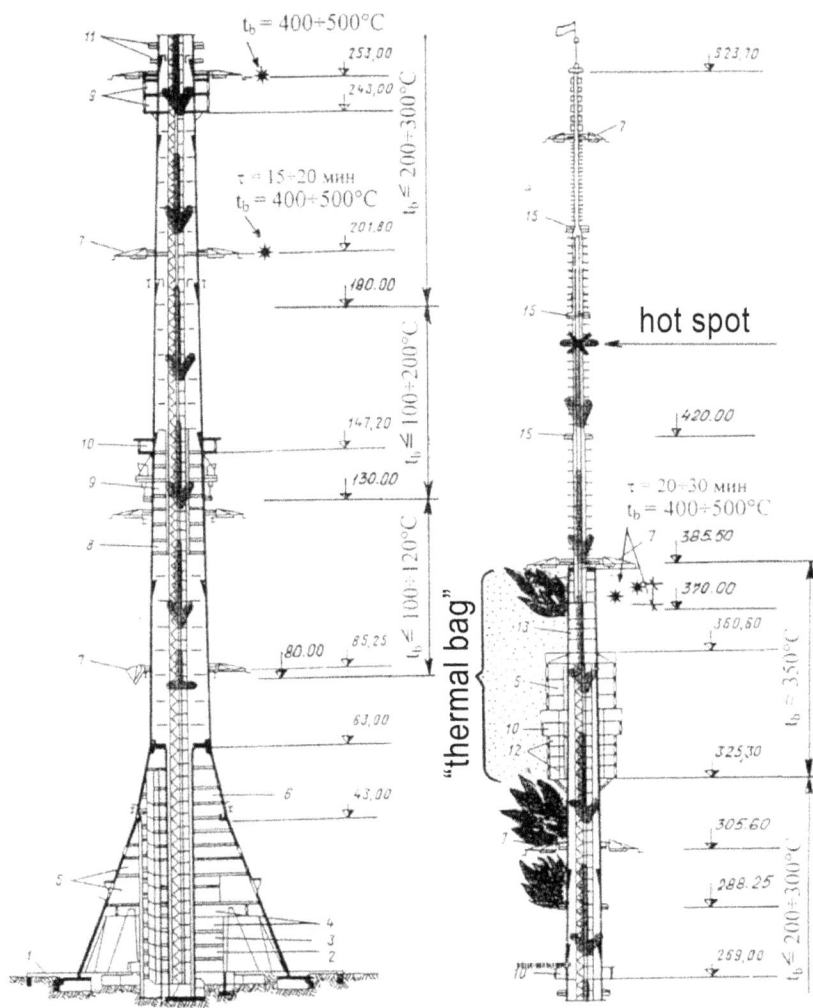

Fig.2.4. Scheme of distribution of a fire inside of the barrel in the Ostankino TV tower: *1* – stylobation; *2* – meeting room; *3* – transformer substation; *4* – technical floor; *5* – transmitting station; *6* – cuisine restaurant; *7, 8* technical premises; *9* – operating room; *10* – a viewing balcony; *11* – balcony to set the receiving antennas; *12* – restaurant; *13* – machine department of elevators; *14* – laboratory; *15* – block-lights.
τ – duration of the effects of the fire; t_b – the temperature on the surface of concrete; → – the direction of propagation of fire

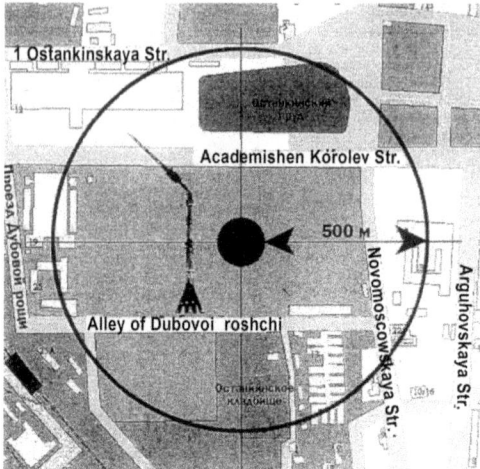

Fig.2.5. The zone of potential falling of debris

Sec.2.3. Engineering Aspects of 09, 11, (WTC, New York)

WTC towers were similar (see Fig.2.6), but not identical. The North tower WTC-1 was 6 feet higher than the WTC-2 and, in addition, it had the 360ft antenna for television and radio broadcasts. In the inside trunk of each tower were 99 elevators and 3 staircases and 16 escalators. The height of each tower was 415 and 417 m. In the plan each tower had a form of a square with the size of the 63,5×63,5 m [2.1-2.7].

The structural scheme of the WTC towers was the trunk in the form of shell. This design is one of the versions of box-shaped structural scheme of high-rise buildings [2.1-2.7]. The out-of-door shells of the WTC towers produced a rigid spatial lattices formed by steel columns of box-shaped section 35, 6×34,3cm and by steel binding beams 132 cm high The binding beams fastened columns of the outer shell at the level of window-sills

Fig.2.6. General view of the WTC towers, NY

on the perimeter of each floor. In general, the external shells of both towers formed a rigid "pipe" fixed on the foundations. Central trunk (core) of the WTC towers was held by 47 steel columns of different shape cross-section.

The design of flours constitutes a spatial system of the steel beams-trusses, connecting with ancillary beams. The supporting beams hold on corrugated steel floor, on which a lightweight concrete plate 100 mm thick was placed. Fire resistance of metal constructions of the towers was supported by the sheeting face of vermiculite plates 3-4,5 sm thick, the thermal evaporation of effective fire retardants, as well as a appliance on the bottom of the floors, suspended ceilings with regulated fire resistance. Metal bearing structures of these towers including fire protection have limits of fire resistance for the loss of the bearing capacity equal to180 minutes.

On September 11, 2001, hijacked by terrorists, aircraft (Boeing 767) battered down WTC towers. The first plane hit the North tower (WTC 1) at 8:45 at the level of 94-98 floors. The second hijacked plane crashed into the South tower (WTC-2) at 9:03, in the area of 78-82 floors.

It should be specially noted that the two high-rise towers of the world trade center after monstrous blow of 180-ton aircrafts, flying at a speed of about 800 km/h, despite the destruction of several tens of load-bearing structures, stood (Fig.2.7). The progressive collapse of the high-rise towers after the planes hit did not happen. The reason was that the survivors bearing structures of the towers after the strike of the aircraft had a sufficient reserve of strength that enabled them to absorb and withstand the additional load from the destroyed structures. Breaking the outer shell of the towers, fragments of aircraft penetrated inside the buildings, damaging and destroying unknown number of bearing columns of the building core. In the premises of the towers (the zone of impact) aviation fuel leaked from the damaged fuel tanks of aircraft, and, as a consequence of this, in the areas of aircraft strikes occurred explosions mixture of spray and vaporized fuel with air (Fig.2.8). This has led to additional destruction and damage of a number of enclosing parts and load-carrying structures of the towers.

Fig.2.7. The destruction of structures of Northern facade of the WTC-1 (North tower) after the strike aircraft September 11, 2001

Fig.2.8. The explosion of the fuel in the zone collision of aircraft with the building of the WTC-2 during the terrorist attacks of September 11, 2001

Thanks to the opening of the glazing, holes in the outer shell of the towers after the impact of the plane, which played the role of explosion-proof buildings (light removable structures), a large number of explosive mixture of fuel with air being thrown out and burned in the environment in the form of a giant "balls of fire" (see Fig.2.8). This reduced the excessive pressure of explosion inside the building to a safe level for the main load-bearing structures and the progressive collapse of the towers did not happen again.

Taking into account the presence in the premises of the WTC towers specific for the offices an amount of combustible materials (fire load of 40 kg/m^2 equivalent wood [2.4]), fires arose in the zone of damage from impacts of aircraft and subsequent explosions in the buildings - a special impact on the structures, associated with development of high temperatures in the heart of fire (Fig.2.9).

Fig.2.9. Tower of the WTC continue to resist to the effects of fire after the impact of the planeand the explosion of fuel

The specificity of the fire in the considered conditions was [2.1-2.6], that:

– fire developed in the rubble of destroyed and damaged structures, objects and things, which were in the offices until the occurrence of an emergency situation;

– the influence of high temperatures of fire happened at survived after the impact and explosion overloaded structures, which took additional load from the destroyed structures;

– when the planes hit the bearing structures of buildings and the subsequent explosion took place, the damage of fire protection occurred on the part of metal structures.

Nevertheless, the WTC tower continued to resist the serial effects of impact, explosion and fire in several tens of minutes. The tower of the WTC-2 resisted to accidental combined effects of 56 min, and the WTC-1 – one-hour 43 minutes and only after that began the progressive collapse of the towers of the WTC as a whole.

It is this amazing survivability of these buildings which allowed to evacuate or to save tens of thousands of people, who have been in the stricken buildings or in close proximity to them.

However, it should be noted that, in spite of the fact that the basic design of the towers had to resist the effects of fire (had limits of fire resistance for the loss of the bearing capacity with consideration for fire protection) of not less than 180 minutes, the loss of robustness of these towers with a CHE like "hit-explosion-fire" occurred much faster. The South tower (WTC-2) has lost its robustness after 56 minutes (Fig.2.10) and the North tower (WTC-1) lost its sustainability through 102 minutes after the fire started.

Fig.2.10. Beginning of the progressive collapse of the WTC-2 (South tower) after 56 minutes of aircraft attack

33

The type of the Pentagon building is an is office building. Floor area is equal to 122600 m^2. Total building area is 613000 m^2. It's a 5 story building designed in the shape of the a pentagon (Fig.2.11). The building inside is divided into blocks, forming 5 concentric rings marked A-E, starting with the inner ring. In the upper three floors the rings of the building are separated by light spaces. Between the second and third rings driveways arranged, known as AE-driveway.

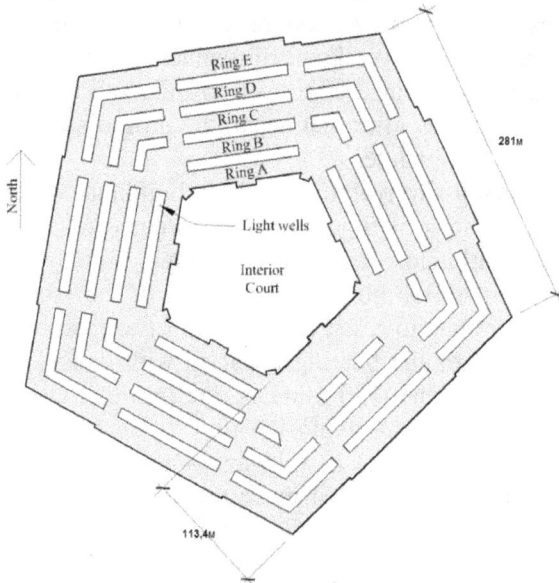

Fig.2.11. General plan of the building of the Pentagon

Before a collision with the building of the Pentagon the Boeing 757 was flying at very low altitude. When it was at a distance of approximately 97 m from the western facade of the building, it was only a few feet from the ground. The blow from the plane occurred on the external façade of the first floor at an angle of about 42 degrees facade (Fig.2.12).

The collision of aircraft with the building was an emergency situation which led to the accident of the type "hit-explosion-fire". The first special effect (the plane) destroyed and damaged a number of structural elements of the first floor of the building. The main attack took over the load-bearing elements of the building – reinforced concrete columns.

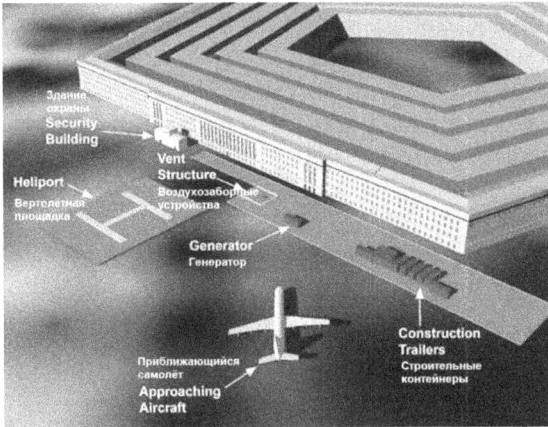

Fig.2.12. The direction of airplane movement before collision

Fragments of the plane penetrated inside the building (see figure 2.13).From the destroyed tanks placed in aircraft wings fuel leaked into the zone of impact inside the building. This has led to the development of the second special impact on the design of the building - the explosion of the air / fuel mixture. Explosion destroyed and damaged another part of the structural elements of the building.

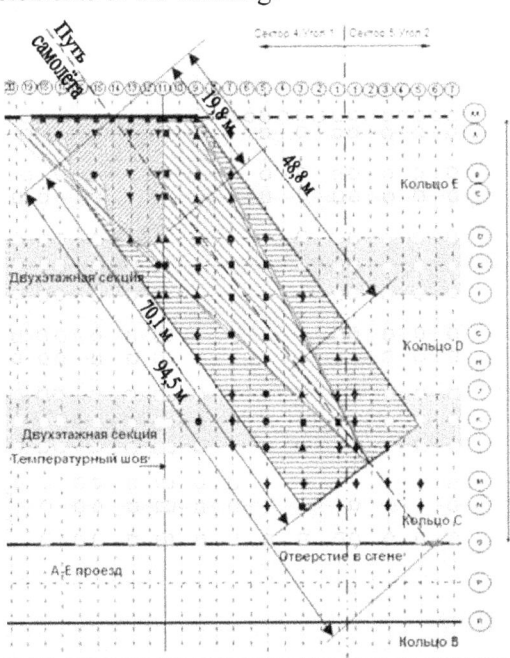

Fig.2.13. Scheme of the damage inPentagon building.

After the impact and explosion inside the building the third special effect, the fire arises and develops in the zone of destruction. The fire covers part of the premises in the path of the aircraft movement. The building of the Pentagon in the first minutes of CHE, in spite of considerable structural damage in the first three rings of the building (Fig.2.14) has retained in general its robustness.

However, after 19 minutes when beginning of the combination of accidental actions like "hit-explosion-fire" took place the progressive collapse of the outer rings of the Pentagon building in the area of "CHE, IEF" (Fig.2.15) happened.

Fig.2.14. View of the facade of the outer rings of the Pentagon building

Fig.2.15. The progressive collapse of the outer rings in the Pentagon building (09.11.2003).

Thus, similar to the behavior of the WTC towers in New York during the events of September 11, 2001, in spite of the fact that the ability to resist the effects of fire of the basic bearing structures in the building of the Pentagon (the limit of fire resistance for the loss of the bearing capacity exceeded 180 minutes), the progressive collapse of structures belong to the outer rings of the building was much faster - in 19 minutes after the start of the terrorist attacks.

Sec.2.5. New Threats to Objects of Construction Complex, Associated with Combination of Accidental Actions with the Participation of Fire

Design of the buildings fire resistance according to the fire safety code permit to provide their robustness in the conditions of fire load during established period of time. Examples of successful resistance of structures exposed to only fire load demonstrate the necessity and effectiveness of such kind of engineering solutions. After the tragic events of September 11, 2001, there were problems related to the safety of unique objects in case of accidental actions with the participation of fire.

One of these problems was the fact that during the terrorist attack at WTC on September 11, 2001 in New York and the Pentagon building these unique objects have lost their robustness faster than it has been regulated by the fire safety codes. In time of these tragic events some previously not considered before hazardous situations took place and their existence lead to premature progressive collapse of the buildings. It is obvious that the appearance of these previously not considered hazardous situations was related to the peculiarities of the combination of accidental situations in comparison with the influence with one fire load only.

The results of the research of engineering aspects on the 09.11.2001 [2.1-2.5] resulted in the assumption, there were any of the following characteristic features in case of combination of accidental actions with the participation of fire, presenting type "impact-explosion-fire" (IEF) appeared:

– There are several groups of structures, with varying degrees of damage.

– As a result of varying degrees of damage these groups of structures will not loose their bearing capacity at the same time, but at different stages of development of the emergency.

– As a result on different stages of emergency serial failure of one invalid groups of supporting structures will increase the load on the remaining structures.

– Increasing the load on the surviving building structures at appropriate stages of the development of the IEF with the participation of fire, will lead to a dangerous result and reduction of the critical temperature of structural heating.

The critical heating temperature of structural material in the event of fire presents such a temperature at which the material loses its ability to resist the effects of fire. A special danger to the buildings of this effect is determined by the obvious consideration that the more there is mechanical load on the structure, the less is the critical temperature of warm-up structures and the faster they lose their bearing capacity in the conditions of the CHE with the participation of fire and the faster comes the progressive collapse of the building as a whole.

For example, the critical temperature of heating of metal columns of the WTC under the influence of fire T_{cr} was only about 500°C. As the research showed [2.1-2.5], the value of the critical temperature of heating of metal columns of the WTC towers in the conditions of the CHE with the participation of fire, reduced to 310-130°C. This effect led to a more rapid loss of the bearing capacity of the WTC towers and premature their progressing collapse.

Thus, the results of the studies testify thatv there exist special danger of combination of accidental actions with participation of fire load and the need to integrate these new dangers and threats while providing complex safety for high-rise buildings and other unique objects.

Sec.2.6. Standardizing Procedure of Structural Protection from the Progressive Collapse

The progressive collapse of buildings refers to the most serious emergencies, leading to heavy casualties and huge material damage. The problem of ensuring the protection of buildings from the progressive collapse considering the terrorist threat is worldwide highly relevant, since the building is one of the most vulnerable objects to such effects [2.1, 2.2].

Research of problems of complex security and antiterrorist protection of unique objects demonstrate [2.9] that one of the most important tasks in this direction is the "improvement of the performance of buildings in slowing or preventing the collapse of the buildings ".

In the formulization of this task two main existing approach to the codification of protection of objects from the progressive collapse were defined:

The first approach: to prevent the progressive collapse of the object as a result of one or another accidental design situation. The second approach: to ensure the resistance of an object in conditions of accidental design situation in a certain period of time prior to its progressive collapse.

The use of an approach to the design of an object is defined, basically, with a kind of accidental design situation. The first approach, in which the progressive collapse of the object is not allowed, is regulated in the code as follow: "Bearing structure must be designed and constructed in such a way that it will be not destroyed by events such as, for example, explosion, impact and consequences of human error to the extent of disproportionate (steep) to the original source".

The use of the second approach is regulated by national and international standards [2.3-3.5] and is required for the design of protection of objects from the effects of fire. This fire load is most often observed cause of the progressive collapse of buildings. Therefore, regulation of protection of buildings from this type of disaster is fundamentally differing from other accidental actions (explosions, impact of transport vehicles, unauthorized alterations, etc.).

In accordance with modern standards [2.3-2.5] for the system of fire protection (SFP) of buildings, the main factor determining the level of safety of the facility in the event of fire shall be adopted by the time (in minutes) of the resistance of the main structures of the object to special effects of fire. For example, in accordance with St. 8 Technical regulation [2.3], "...the Building or other structures shall be designed and constructed in such a way that in the course of operation of a building or structure in the case of fire they will meet the following requirements: – the preservation of the stability of a building or structure, as well as the strength of bearing structures of the time necessary for the evacuation of people and perform other actions, aimed at reducing the losses from the fire...".

Similar requirements exist in the Eurocodes:

– "Fire design presents the design of the bearing structure to ensure the required performance in the time of the fire". "In case of fire, the structural strength should be adequate for the required considerations (fire protection) period of time".

– The resistance of an object to the occurrence of the limit state under the influence of fire is the international fire-technical characteristics, and standardized as a special measure, which got the name of "fire- resistance".

STO [2.5] regulate, for example, the resistance of the basic structures of high-rise buildings (above 75 m) exposed to fire (their fire- resistance) with value of 180 minutes. This means a high-rise building for three hours should not lose its resistance or geometric invariability after the occurrence of a fire.

That is, in principle, allow for the possibility of the progressive collapse of the object after the expiration of the time and with considering these circumstances in standardizing other elements of the system of fire protection. The combined effect of them in the conditions of accidental situations must ensure the safety of people, property, fire-fighting, etc.

Even more urgent is the need to estimate the time of the resistance of objects before the start of the progressive collapse. It is necessary in the design to protect unique objects from the progressive collapse in the event of combined accidental actions with the participation of fire [2.8-2.9]. Thus, the main criterion that determines the choice of the first or the second approach to the design of protection of the object from the progressive collapse, is the presence or absence in the codes or standards some obligatory requirements. These requirements will concern the time estimation of resistance of an object before the progressive collapse in the conditions of the accidental design situation.

Sec.2.7. Protection of Structures from Progressive Collapse. (First Approach)

There are a number of methods of assessment of the "vulnerability", "survivability", "sustainability" and protection of structural systems in different variants of danger, the purpose of which is to exclude the possibility of their progressing collapse [2.1, 2.2].

In the Recommendations [2.1] for the protection of high-rise objects from the progressive collapse the first approach is used- to prevent the progressive collapse of an object as a result of one or another accidental design situation.

These Recommendations [2.1] are intended for the design on resistance against the progressive collapse and construction of new, reconstruction and verification of the constructed high-rise buildings of any structural systems of a height of more than 25 floors (75 m) for such accidental actions as: fire, explosion, impact of transport vehicles, unauthorized alterations, etc.

In the case of these accidental actions local destruction of separate vertical bearing elements in the same floor or overlap area of one floor are allowed. Moreover, "this initial destruction should not lead to the collapse and destruction of structures, to which the load is transferred. This load was perceived by the elements which were damaged previously from accidental actions " [2.1].

To achieve this goal it is proposed to make the design of the high-rise buildings on the resistance against the progressive collapse [2.1]. As a local (hypothetical) destruction these Recommendations consider damage (destruction) of vertical structures of one (any) floor of the building, the limited range of up to 80 m^2 (diameter 10 m) for buildings height up to 200 m and up to 100 m^2 (diameter of 11.5 m) for buildings above 200 m.

Sec.2.8. Protection of Structures from Progressive Collapse. (Second Approach)

This approach, as it has already been noted above, is required for the design of protection of objects from the effects of fire. In the national standards [2.3-2.5] the index, which characterizes the resistance of building structures exposed to fire, is called "the limit of fire- resistance", and the indicator, which characterizes the ability of the building as a whole resist the effects of fire, is called "the degree of fire- resistance". With the help of these indicators [2.3-2.7] time is regulated during which the structures of buildings have to resist high temperature of the action of fire before the onset of the progressive collapse. Design of the fire- resistance of buildings is a necessary element of ensuring their protection from the progressive collapse in the conditions of fire.

The essence of the design of the fire resistance of buildings consists in the following procedures:

- Estimation of the moments of time, on the expiration of which in the conditions of fire the main designing structures reaches the rate of limit states, including the loss of the bearing capacity. This procedure has received the name - estimation of actual values of the limits of fire resistance of the main structures.

- Verification of the conditions for the protection of the object from the progressive collapse in the conditions of fire. If the values obtained for the actual limits of fire- resistance of the basic structures correspond to (equal to or higher than) the minimum possible (required by standards) values of the fire- resistance limits of these same structures, the protection for the object from the progressive collapse in the event of fire will be achieved [2.7].

In this kind of design data are used about the actual limits of fire- resistance obtained from catalogues or as a result of analysis of building structures for fire resistance [2.7]. Design of building structures for fire resistance is one of the aspects of the analysis of structures for limit states of [2.7]. A remarkable case of the correct design of the fire resistance of an object ensuring the protection of the progressive collapse is the behavior of the Ostankino TV tower during a fire on August, 2000 [2.6].

Sec.2.9. Analysis of Different Approaches to Protection of Objects from Progressive Collapse

The practice of international and national standards to protect buildings from the progressive collapse testifies the fact that the application of the above-considered the first or the second approach in the solution of such kind of structural design depends on the type accidental design situation.

In this connection there is a need to clarify the list of beyond design accidental situations on which enlarge upon the application of the first approach.

It is necessary to exclude fire from the list of beyond design accidental situations (fires, explosions, karts sinkholes, defects of structures and

41

materials, incompetent reconstruction (redevelopment), which are considered in [2.1, 2.2].

This necessity is conditioned by the following circumstances:

1. For the design of protection from the progressive collapse of the building in the conditions of fire it is necessary to use the second approach, associated with the need to measure the time of the building resistance before the onset of the progressive collapse. The methodology presented in [2.1, 2.2], does not allow to make such estimation.

2. In the conditions of fire the observed scale of local damages of structures may be really many times higher than the scale of local injuries, taken into account in the Manual [2.1].

The fact is, that the fire in buildings may develop within the so-called "fire compartment". And allowable area of fire compartments, according to the codes of fire safety [2.6, 2.7], exceeds thousand square meters. This means that, in the event of fire, the area of local failures may exceed thousand square meters, which does not fit in the possibilities of the methodology of [2.1].

In connection with these circumstances there is also a need to clarify the terminology using in this field of the science and practice. Concept of "vulnerability", "sustainability", "survivability" describes a property of structures to withstand beyond design situations without the occurrence of the damage, which would be out of proportion to reason that caused the damage. This means that these concepts in their content and meaning not allow operating with the time of the resistance of an object in the conditions of fire and belonging to the one described above the first approach to the decision of problems of protection of objects from the progressive collapse.

For the purposes of assessing the time of resistance of objects before the onset of the progressive collapse in national and international practice such terms are of the use as "durability", "fire- resistance, "robustness "and others. [2.3-2.9].

Accordingly, in the national and international design practice special design methods of protection of buildings from the progressive collapse for the case of a fire have been developed and applied [2.3-2.7]. These special methods are called - the design of the fire resistance of buildings.

The design of the protection of high-rise and unique objects from the progressive collapse relates also to this class of problems due to combination of accidental actions with the participation of fire [2.3-2.6]. Below are the general approaches and methods of estimating the resistance of buildings against the progressive collapse in accidental situations with the participation of fire [2.6-2.9].

Sec.2.10. Theory of the Fire Resistance of Building Structures

The special nature of the hazards of exposure to fire in buildings is confirmed by the fact that in the international standards for fire safety of buildings and structures introduced special indicators, characterizing the ability of objects resist the effects of fire, and in these terms the main characteristic of providing the security of the facilities in these circumstances is the "time" resistance of objects of the effects of fire.

In the national codes of practice [2.3-2.5] index, which characterizes the ability of building structures resist the effects of fire, is called "the limit of fire resistance", and the indicator, which characterizes the ability of the building as a whole resist the effects of fire, is called "the degree of fire resistance".

With the help of these indicators the time is regulated during which buildings and structures must resist high temperature in case of fire.

The need to solve a complex of scientific and engineering problems for the assessment of the fire resistance of buildings and structures have stimulated the development of international studies, the results of which were formed in the theory of fire-resistance [2.6].

The essence of analysis of fire resistance is to specify the time after which the structure in conditions of the impact of fire will lose their bearing capacity or the heat-insulating ability (Fig.2.16).

Design of building structures for fire resistance is one of the varieties of limit state analysis of structures.

For example, in the design of building structures for fire resistance factor γT is used as "the working conditions of structural material in the event of fire" [2.6]. This factor takes into account the peculiarities of the change in resistance of conventional building materials by heating in the conditions of fire.

In accordance with this approach, the working conditions factor for regular structural materials in the event of fire, represent the relationship between resistance of materials R (T) and a temperature of their warm-up "T".

The dependencies presents in relative form as:

$$\gamma T = R\ (T)\ /\ R = f\ (T), \tag{2.1}$$

where R (T) is the strength of a material at temperature T;
R – initial strength of the material.

Relations of the type (2.1) have been obtained for all of the basic structural materials as a result of years of special experimental research. These studies show that under heating in the conditions of fire the resistance of regular structural materials after a certain temperature begins to decrease rapidly.

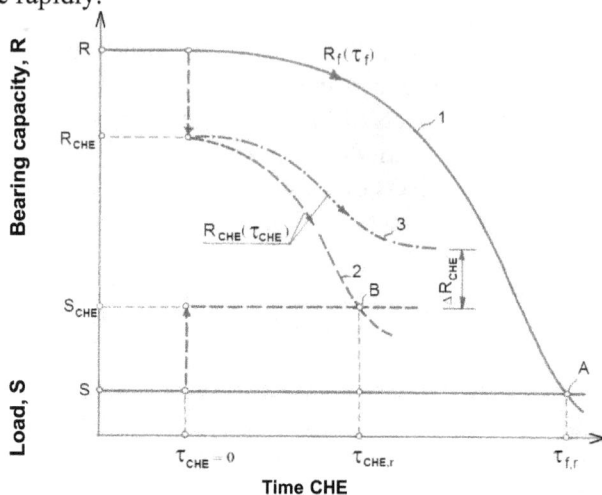

Fig.2.16. General scheme for decision of the problems of assessment of the resistance of buildings structures in accidental situations with the participation of fire: 1 – assessment of the fire resistance of structures (with the loss of the bearing capacity $R_f(\tau_f)$ in time τ_f of fire (t.A. on a curve 1); 2 – estimation of resistance of the structures () in case of combination of special actions- "hit-explosion-fire" (CHE IEF) (loss of bearing capacity (t.B. on a curve 2); 3 – change of bearing capacity of structure R_{CHE} as CHE IEF that does not lead to the loss of the bearing capacity and preserve some residual provision of its strength (ΔR_{CHE})

These dependencies are used, at the present time, as reference data in the design of building structures for fire resistance.

Fire resistance of building structures in a number of cases conveniently assessed using the indicator of "critical temperature" warm-up materials of structures in the conditions of fire.

The concept of "critical temperature" warm-up materials of structures is one of the basic indicators used in the theory of analysis of building structures for fire resistance [2.6].

The main feature of the theory of the fire resistance of building structures is the fact that the value of the critical temperature of heating of a structural material in the assessment of the limits of fire resistance has a fixed value corresponding to the design level of the representative working load on this structure.

44

Sec.2.11. Resistance of Buildings Structures due to Combination of Accidental Loads (CHE) with the Participation of Fire

Above was shown the need to solve new problems - provision of the necessary resistance to unique objects with different combination of accidental actions (CHE) with participation with fire.

The search for the solution of this problem [2.7-2.9] has identified the need for introducing a special concept which, in its physical sense would reflect the ability of an object to resist CHE with the participation of fire within a certain time.

The closest prototype of such kind of the concept is the notion of fire resistance of building structures.

Proven in [2.5,2.6] the belonging of methodical and physical principles underlying in the concepts of "durability ", "fire-resistance", "resistance" of objects allowed to offer a more general notion of "resistance of objects due to combination of accidental actions (CHE) with the participation of fire", which you can use at the decision of a wide class of problems.

With these considerations in mind:

• Resistance of structures in case of CHE with the participation of fire $(\tau che, r)$ is the time during which the structure retains its carrying and non-bearing functions in the conditions of CHE with the participation of fire.

• Resistance of the building to CHE with the participation of fire (Dcне, r) is the time during which the building as a whole resists to the effects of accidental CHE without buckling and geometric immutability. The resistance of the buildings in a CHE is determined by the resistance of its main structures in these conditions.

• Fire resistance is the time, during which the structure retains its bearing and non-bearing function in the conditions of the influence of the all loads and high temperature of the fire.

The essence of the proposed method for estimation of the buildings resistance in a CHE with the participation of fire is in the analysis of the change in time of the carrying capacity for characteristic groups of structural elements and loads on them at a given scenario of CHE with the participation of fire, with consideration also the peculiarities of the behavior of structural materials in the considered conditions (curves 2 and 3 in Fig.2.16).

Taking the above into account the progressive collapse of the object is the last avalanche effect stage of developing in time of the process of the serial depletion of resistance of structural elements, which leads to a loss of the overall sustainability of the object as a whole.

All the diversity of the actions of a CHE in the building is offered to express through the consequences of these actions on the status on the "key" of the structural elements of the building, which determine its resistance against the progressive collapse.

"Key" design elements of a building with a CHE participate and play a decisive role in ensuring the general resistance and geometric immutability of the building in the considered conditions. As a "key" of the structural elements on the basis of structural schemes of modern buildings bearing elements like columns, bearing walls, frames, etc. can be considered.

All of the "key" structural elements in the examined building are divided into several specific groups depending on their status in the conditions of a CHE and the ability of these elements to resist CHE.

Taking into account the different resistance of typical groups of "key" elements the process of exhaustion of the resistance of the buildings in given CHE scenario will take place in some stages, as a result of consistent loss of the resistance of the various groups of "key" elements.

The loss of resistance of a particular characteristic group of the "key" elements of the building on one or another of the design stage of CHE leads to an increase of overloading for the remaining group of survivors of the "key" elements. This, in turn, leads to a deterioration in their conditions of work (reduces the structural strength, reduces the critical temperature of the heating in the event of fire, etc.) and leads to decreasing their resistance in case of CHE.

The presence of these processes leads to the necessity of consideration of a number of design stages in the development of CHE. Each of the isolated design stages of a CHE development will correspond to a loss of resistance for a certain characteristic group of the "key" elements.

The resistance of the building structure for combined accidental actions determined by the time τ_{CHE} from the beginning of the combination of accidental actions (CHE) until the moment when the carrying capacity of the structure R_{CHE} is not enough for the perception of the load S_{CHE}, annexed to it at various stages of CHE.

Fig.2.16 shows the overall scheme of estimating of structural resistance in case of CHE.

Resistance of a structure in CHE ($\tau_{CHE, r}$) is determined by analysis of the change of its carrying capacity (R_{CHE}) and annexed to it load (S_{CHE}) at various stages of CHE, in accordance with the scenario of "CHE IEF".

Resistance of a structure in a CHE with the participation of fire $\tau_{CHE, r}$ is determined by the expression:

$$\text{If } R_{CHE}(\tau_{CHE}) < S_{CHE}, \text{ then } \quad \tau_{CHE} = \tau_{CHE,r} \qquad (2.2)$$

The entire building has exhausted its resistance against the progressive ccollapse ($D_{CHE,\,r}$) and will lose resistance in the selected scenario CHE with the participation of fire, if the entire characteristics group of the "key" elements comes to limit state for the loss of the bearing capacity at any of the stages of CHE.

Time τ_{CHE} will determine the actual resistance of the building against the progressive destruction ($D^{act}_{CHE,\,r}$) for the selected script of CHE. The value of the resistance of the building in these conditions can also be determined from the relation:

$$\text{If}\quad \tau_{CHE} > \tau_{CHE,r} \qquad\qquad \text{then}\quad \tau_{CHE} = D_{CHE,r} \qquad (2.3)$$
$$\text{(for all "key" elements)}$$

The building will retain a certain part of their resistance against the progressive collapse and will not be completely destroyed when a given scenario of a CHE. It will be so if certain groups of "key" structural elements of the building will not reached the limit state for the loss of the bearing capacity after consideration of all design stages of CHE. In this case the building will retain its integrity, but will receive a certain level of damage.

The ability to save the building of its integrity in the given scenario of the CHE is determined from the condition:

$$\text{If}\quad \tau_{CHE} < \tau_{CHE,r} \qquad\qquad \text{then}\quad D_{CHE,r} > \tau_{CHE} \qquad (2.4)$$
$$\text{(for all "key" elements)}$$

The proposed method can be used to address two types of tasks.

Task I (direct proposition). Estimation of resistance of the building under various scenarios of the CHE with the participation of fire.

Task II (inverse proposition). The definition of the allowable number of "key" building structures, which can be destroyed or damaged in a CHE with the participation of the fire, on the basis of a given (codified) resistance of buildings against the progressive collapse ($D^{req}_{CHE,r}$). Codified level of resistance of the building is determined on the basis of the acceptable levels of risk, the safety of the people and the preservation of the building.

The proposed theory and methods for estimation of resistance of buildings give an opportunity to any level of detail of the original model of the building and the design scenario of development of the combined accidental actions with the participation of fire.

Ch.3. Natural Disasters and Structural Robustness

Sec.3.1. Disaster Prediction Problems

The term "disaster" is understood here as any environmental change endangering human lives and materially deteriorating living conditions. A considerable part of disasters comprises natural calamities. Disasters can originate both inside the Earth due to tectonic processes (earthquakes, volcanic eruptions) and near or on its surface due to atmospheric processes (floods, tsunamis, hurricanes, tornados, land and mud sliding, avalanches, karsts, ground heaves and settlements). In many cases successions of interdependent disasters are possible, including those occurring in different media (earthquake-tsunami, earthquake-landslide, and hurricane-flood, etc.). Analysis of conditions associated with the onset and the development of dangerous natural processes is the subject of both natural research and engineering study. The mechanisms of dangerous natural phenomena can be described by direct cause–effect relations. Predicting the type, time and magnitude of an expected disaster, if feasible, can only be probabilistic. Therefore, using a probabilistic approach and a reliability theory appears to be the most efficient and the only practical tool for analyzing structures operating in areas where natural calamities can be expected.

The level of understanding the origination of natural disasters and, hence, the efficiency in predicting their time, conditions and severity lags behind the practical needs of national economy. And so does the development of measures for their prevention and mitigation. This is partly due to the absence of common approaches for modeling some natural disasters and methods of their prediction. The forecast of rare disastrous events can be based either on physical models of the phenomenon in question, or using statistical methods, of both. For the latter, statistical information for a rather long period of time is needed. Practically, however, the prediction is usually based on limited information, often imprecise and sometimes merely incorrect.

The accuracy of statistically-based forecasts fluctuates within a certain range.

Different methods should also be used in prediction. The methods commonly used for sufficiently substantiated statistically-based predictions [3.1-3.6] include: multi-dimensional regression analysis, theory of

---

[*] Reliability Engineering Consulting, San Diego, USA

48

quantitative analysis, graph theory for error analysis, Delphi method (method of expert evaluation), and statistical analysis. In addition to these five approaches, forecasting the disastrous events and preventing the maximum risk and losses due to abnormal actions in recent decades has been based on the theory of fuzzy sets [3.7]. It's application is due to the fact that any classification, any rule of decision making, any model (theoretical or calculated) can be correlated with its fuzzy analogue. For example, classification implies the breakdown of a multitude of elements into classes or groups of similar elements. A prescriptive classification refers each element to a single definite class, whereas in a fuzzy classification each element can belong to different classes depending on certain conditions. The fuzzy classification is generally more realistic than the prescriptive one. The use of the theory of fuzzy sets makes it possible to develop optimal solutions in accordance with practical restrictions.

Sec.3.2. Statistical Evaluation of Natural Disasters

Unlike traditional methods, the probabilistic approach makes it possible to evaluate a possible magnitude of disastrous actions in a particular area when analyzing both structures and soils. Therefore, when elaborating a probabilistic concept for natural disasters one should primarily consider in a general form the feasibility of using the statistical approach for representing the disastrous effects.

The statistical analysis for the problem should be focused on a probabilistic prediction of the time, location and magnitude of a natural disaster, or alternatively, on the occurrence probability of a particular type of disaster at the given service life and location of the structure during the given time period. Besides probabilistic predictions, a direct forecasting based on warning signs can also be used. Reliable warning signs, however, are often detected just before the disaster and cannot be taken into account in a long-term prediction influencing the engineering solution.

Statistical methods are used to have a prior notion of the frequency and extent of disasters possible in a particular region. Observations for previous years can give the information on frequency and other parameters of the past disasters. The future characteristics can be estimated by extrapolating the probability of the past events. The estimate, however, can be rather conventional since the acquired statistical data refer to a limited time range. Therefore, the data processing should be based on specially developed statistical models whose physical adequacy to the phenomena under consideration would make the extrapolation trustworthy. Since natural disasters are extreme occurrences (earthquake or/and tsunami of high intensity, etc.), they can be characterized by the "statistics of rare phenomena". Extreme values are usually associated with small occurrence probabilities. The Poisson's distribution appears to be most appropriate in this case for modeling the time character of disasters by the Poisson's process.

This law determines the number of occurrences of rare events, while the theory of extreme values [3.8] considers their values. For random events where extreme values play major role, the asymptotic theory of extreme order statistics provides in some cases relatively accurate but mostly approximate probabilistic models. Therefore, if the basic assumptions of the model have a similarity to the main conditions of a real situation of a catastrophic action, the complex real conditions can be simulated by a considerably simple asymptotic model. Areas, where dangerous phenomena can occur at intensity levels exceeding those on the record (earthquake exceeding the design level, etc.), can be determined and assessed by the test's observations of similar but less intensive occurrences.

A specific feature of natural disasters (as well as human-caused disasters) is that they are virtually unavoidable. Natural disasters are characterized by power and uncontrollability. Typical of human-caused events is that they result from quick development of modern technologies and their products. These factors and the human factor constitute a weak link in a chain of events leading to tragic consequences (Chernobyl, for example). The main task here is to predict possible disasters, localize them and mitigate possible losses. Design of any structure should be preceded by the analysis of all possible types of natural or human-caused disasters in terms of the occurrence probability, secondary disalocalization, preventive measures not connected with the design methods, and the possible damage.

Sec.3.3. Safety Criteria of Unique Structures

Prior to discussing the safety criteria we should clarify the notion of a unique structure and natural or other effects that, determining its vulnerability, are detrimental for human health. The notion of the structural uniqueness and that of the danger of natural or other calamities are interconnected. Considering the structural safety in terms of the danger to human life and health, we should not connect the uniqueness of the structure with its cost or with the expected material losses. The uniqueness should as well be linked with the level of the danger to people, regardless of its probability and of factors causing it, such as the structure's operational profile, size, how it has been built, the presence of radioactive products, etc. Hence, unique structures are those whose damage or collapse, no matter how low its probability could be, endangers the human life and health either inside or, what is more often, outside the structure.

The foregoing definition of structural uniqueness allows one to refer to such structural projects of national economy (energy, transportation, and others) including social sphere, whose damage and collapse would endanger human life and health. Vulnerability of unique structures exposed to disastrous actions and possibility of their damage or collapse depend on:

• the extent to which the loads due to disastrous events exceed standard loads;

• the influence of secondary factors (explosions, fires) due to disastrous events;

• the errors occurred in the design, analysis and in choosing the location of the structure, both during its construction and maintenance;

• poor workmanship, inadequate structural materials, and outdated standards for materials, construction and maintenance.

When analyzing structural vulnerability or safety it is expedient to single out critical elements on which the structural safety depends most. For many structures these are the load-bearing members that determine the structural strength and stability (foundation, columns, floors, joints, supports, etc.), as well as the members resisting explosion or fire caused by disastrous events and ensuring reliable operation of safety systems. For a number of unique buildings the critical elements are associated with the protection from radioactivityand with radiation safety.

Different characters of the critical elements require performing, when choosing safety criteria of unique units, a systematic analysis in order to identify these elements and to assess the consequences of their failure. The systematic analysis of structural safety should include elaborating the critical scenarios of the event in question taking into account its possible development, the structure's uniqueness, the presence and type of the critical elements, the consequences of their failure, the implications for human life and for the environment, etc. Generally, every natural phenomenon and every unique structure requires a scenario allowing one to take their specificity into account and to obtain statistical data for generalizing the consequences. The elaboration and analysis of the scenarios require a great professional skill and effort of people acquainted with the task.

To specify qualitative and quantitative safety criteria of unique structures exposed to any types of natural effects, an integrated approach should be based on:

51

• systematic deterministic analysis of scenarios describing how the disastrous event affects the unique structure revealing particular quality criteria;

• probabilistic risk analysis determining particular and general probabilistic safety criteria that include those for the limit states representing the extent of failure, and criteria for endangering human life and health;

• cost-benefit analysis for assessing the safety based on optimization of investments for protection against unfavorable effects with due regard for socio-economic factors.

Sec.3.4. Robustness of Structural Systems

Different situations that had not been foreseen during the design of a structure can occur as a result of natural or human-caused abnormal actions. These situations can be classed according to failure type, degree of damage, and final state. The following types of failure can be considered for the ultimate limit state:

• loss of strength in time of plastic, brittle, or fatigue failure of elements;

• elastic or inelastic buckling of structural elements;

• loss of the stable equilibrium of the whole structure.

According to the degree of its intensity the damage can include the following.

– Rapidly progressive failure of the whole structure. This form of failure is typical for brittle failure when a damage of separate elements can cause dynamic effects in other elements of a structure [3.10].

– Gradually growing failure of an accidental character as a result of plastic deformations. This situation will stop operation and require repair. This form of failure is typical for structures made of ductile materials when a failure of separate element(s) is accompanied by growing displacements and redistributions of inner forces.

It should be noted that in failure analysis of structures, the failure process has an avalanche-like character of a sequence of failures of elements of a multi-member system. And the notion of damage can be applied to both partial failure and complete collapse of the system. In a majority of cases, however, a number of individual (or partial) failures do not necessarily leads to a total breakdown. This is usually due to a redistribution of stresses which typically occurs in statically indeterminate systems, so the structure continues to operate though at a lower efficiency and redundancy. The situation can be characterized by the reserves of load-bearing capacity of the structures. The safety margins are usually implicitly embedded in the design, based on expe-

52

rience or engineering judgment. To provide a reasonable reliability level the structure should be designed in such a way that the zone, where damage to main load-bearing elements is likely to occur, should be localized and isolated [3.9].

The term robustness of a structural system is understood as the ability to operate in the period of disastrous events and to prevent progressive failures. Robustness is an important and, especially for a unique structure, indispensable property, since its reliable performance is possible only if an appropriate level of survivability is ensured. It appears natural that a quantitative assessment of this ability should be based on a probabilistic approach, similar to what has been conventional for structural reliability evaluation.

Following this logic, the level of robustness can be determined by a probability of certain events characterizing the process of failure [3.10]. It is logical to consider how a critical state is attained in the process of successive failures of members. This can be the failure of a number of members and formation of an instantaneous mechanism, or the failure of isolated members, etc. According to this approach, a structure, having been in a damaged condition, can be considered capable to survive if the probability of that event is not so high, as compared to its undamaged condition (other criteria can be used as well).

Method of approaching the discussed problem consists of introducing the robustness index in the following form:

$$\eta = \frac{P_f}{P_f'} \tag{3.1}$$

Where P_f is the probability of failure of the designed system; P_f' – probability of failure of the same system when some members failed. Robustness factors η are in [0, 1] interval. The greater the value of η , the greater is the reserve of robustness in structural system [3.11]. This is illustrated in the example of a steel frame in Fig.3.1.

The horizontal girders have a span of l=6 m. The height of vertical columns $h = 4$m.

All members of the frame are made of I-beams with section modulus $W = 6.15 \cdot 10^{-5}$ m^3 for the 1st floor column, $W = 8.28 \cdot 10^{-5}$ m^3 for the 2nd floor column, $W = 1.270 \cdot 10^{-4}$ m^3 for the 1st floor girder, $W = 1.098 \cdot 10^{-4}$ m^3 for the 2nd floor girder. Probabilistic analysis was performed taking into account random nature of applied loads and yield stress of frame material, with the given distribution probabilities. Table 3.1 contains parameters of these distributions. Calculations were made using linear programming method (simplex method) with the application of the direct integration of distribution function [3.12]. Probability of failure is $P_f = 5.51 \cdot 10^{-5}$.

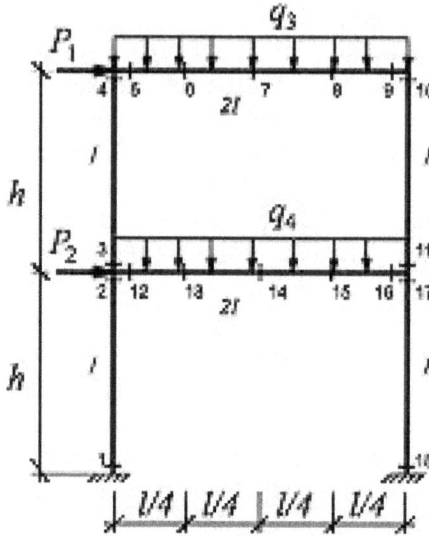

Fig.3.1. Two-story frame

Table 3.1

Random value	Distri-bution	Mean value	Standard deviation s	Parameters of distribution	Design values
Wind load P_1, P_2	Gumbel	$0.144\,kH/m^2$	$0.037\,kH/m^2$	$u = 0.127\,kH/m^2$ $z = 0.029\,kH/m^2$	$0.2576\,kH/m^2$
Snow load q_3	Gumbel	$1.1418\,kH/m^2$	$0.4681\,kH/m^2$	$u = 0.931\,kH/m^2$ $z = 0.365\,kH/m^2$	$1.6\,kH/m^2$
Load due to use q_4	Gauss	$0.88\,kH/m^2$	$0.21\,kH/m^2$	–	$1.68\,kH/m^2$
Yield point σ_y	Weibul	$305.25\,M\Pi a$	$25\,M\Pi a$	$\beta = 14.3$ $\alpha = 316.42\,M\Pi a$ $x_0 = 0$	$245\,M\Pi a$

54

Table 3.2

№ sec-tions	Probability of failure P_f					
	Lowering of aria moments W in different sections					
	5%	10%	25%	50%	75%	95%
1	$5.51 \cdot 10^{-5}$	$5.51 \cdot 10^{-5}$	$5.51 \cdot 10^{-5}$	$5.51 \cdot 10^{-5}$	$7.53 \cdot 10^{-5}$	$8.42 \cdot 10^{-5}$
2	$5.51 \cdot 10^{-5}$	$5.51 \cdot 10^{-5}$	$5.51 \cdot 10^{-5}$	$5.51 \cdot 10^{-5}$	$7.41 \cdot 10^{-5}$	$8.94 \cdot 10^{-5}$
3	$5.51 \cdot 10^{-5}$	$5.51 \cdot 10^{-5}$	$5.51 \cdot 10^{-5}$	$5.51 \cdot 10^{-5}$	$5.51 \cdot 10^{-5}$	$5.51 \cdot 10^{-5}$
4	$5.83 \cdot 10^{-5}$	$5.96 \cdot 10^{-5}$	0.000101	0.000207	0.000389	0.000570
5	$5.51 \cdot 10^{-5}$	$5.51 \cdot 10^{-5}$	$8.42 \cdot 10^{-5}$	0.000122	0.000309	0.000547
6	$5.51 \cdot 10^{-5}$	$5.51 \cdot 10^{-5}$	$5.51 \cdot 10^{-5}$	0.000107	0.000755	0.004562
7	$6.19 \cdot 10^{-5}$	$7.90 \cdot 10^{-5}$	0.000303	0.001246	0.006322	0.025580
8	$5.51 \cdot 10^{-5}$	$5.51 \cdot 10^{-5}$	$5.51 \cdot 10^{-5}$	$8.34 \cdot 10^{-5}$	0.000734	0.004771
9	$5.51 \cdot 10^{-5}$	$5.51 \cdot 10^{-5}$	$5.51 \cdot 10^{-5}$	0.000137	0.000319	0.000593
10	$5.95 \cdot 10^{-5}$	$6.86 \cdot 10^{-5}$	0.000103	0.000207	0.000392	0.000564
11	$5.51 \cdot 10^{-5}$	$5.51 \cdot 10^{-5}$	$5.51 \cdot 10^{-5}$	$5.51 \cdot 10^{-5}$	$5.51 \cdot 10^{-5}$	$5.51 \cdot 10^{-5}$
12	$5.51 \cdot 10^{-5}$	$5.51 \cdot 10^{-5}$	$5.51 \cdot 10^{-5}$	$5.51 \cdot 10^{-5}$	0.000112	0.000265
13	$5.51 \cdot 10^{-5}$	$5.51 \cdot 10^{-5}$	$5.51 \cdot 10^{-5}$	$5.51 \cdot 10^{-5}$	0.000224	0.000873
14	$5.51 \cdot 10^{-5}$	$5.51 \cdot 10^{-5}$	$5.51 \cdot 10^{-5}$	0.000890	0.001063	0.002327
15	$5.51 \cdot 10^{-5}$	$5.51 \cdot 10^{-5}$	$5.51 \cdot 10^{-5}$	$5.51 \cdot 10^{-5}$	0.000229	0.000871
16	$5.51 \cdot 10^{-5}$	$5.51 \cdot 10^{-5}$	$5.51 \cdot 10^{-5}$	$5.51 \cdot 10^{-5}$	0.000112	0.000259
17	$5.51 \cdot 10^{-5}$	$5.51 \cdot 10^{-5}$	$5.51 \cdot 10^{-5}$	$5.51 \cdot 10^{-5}$	$7.30 \cdot 10^{-5}$	$8.27 \cdot 10^{-5}$
18	$5.51 \cdot 10^{-5}$	$5.51 \cdot 10^{-5}$	$5.51 \cdot 10^{-5}$	$5.51 \cdot 10^{-5}$	$7.34 \cdot 10^{-5}$	$8.33 \cdot 10^{-5}$

Table 3.2 shows that in the case of a failure of any cross-section, probability of failure for frame will not exceed the value $P_f' = 0.02558$ (the failure of cross-section 7). The failure of cross-section 7 will not lead to the collapse of all structure but essentially decreases its survivability. Even the full failure of cross-sections 2 or 11 has no influence on probability of this frame. The failure of the cross-section 1, 2, 17 or 18 has also no essential influence at this probability. Robustness index of the considered frame with regard to the failure of cross-section 7 constitutes:

$$\eta = \frac{5.51 \cdot 10^{-5}}{0.02558} = 0.00215$$

If in the process of structure exploiting some actions will be ensuring, then the probability of the failure of the whole frame in case when one cross-section failed, can be decreased to the value $P_f' = 0.004771$. Survivability index will be:

$$\eta = \frac{5.51 \cdot 10^{-5}}{0.004771} = 0.0115$$

At Fig.3.2 graphs due to dependences between probability of failure and weakening of cross-sections 7, 8 and 3 are presented.

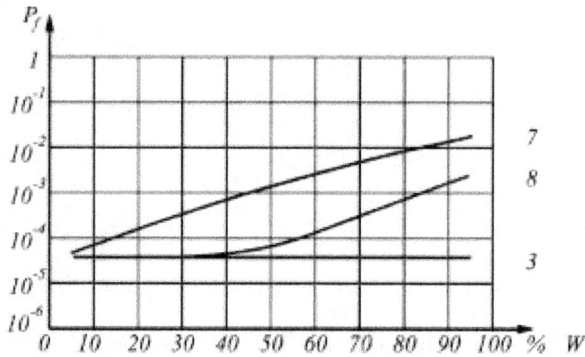

Fig.3.2 Dependence between P_f and W

Ch.4. Dynamic Actions Generating Structural Failure

Kulyabko V.V.[*]

Sec.4.1. General Comments

This work devoted by analogy with work [4.2], to the "harmful" and "useful" dynamics of structures. It is advisable to consider first and if possible, categorize themselves kinds of dynamic loads and actions, linking them together with the reasons of structural destruction (or condition close to the collapse, or the discomfort of the staff and tenants). Then it is recommended to consider the main weaknesses and ways to improve the design on dynamic loads. And, finally, to offer recommendations theoretical, experimental and organizational character directed on reduction and prevention of accidents for various reasons.

First of all, let us pay attention to the fact, that the dynamic load poorly investigated (indirect methods of accounting of the dynamic loads do not reflect their essence and therefore errors inherent in indirect substitution may also be the reason of accidents!). However, in recent years, after writing the fundamental works and the creation of standards [4.3-4.5], a lot of absolutely new results appeared!

Various static-dynamic interactions of building structures poorly studied and are not always correctly accounted in design (and, consequently, may also be the causes of accidents), for example:

- with underground bases and its inclusions (karsts and other emptiness, rocks, boulders, manifolds, etc.);
- with the flow of air masses and wind load, when this interaction causes complex aeroelastic oscillations, including powerful self-oscillations of the type "Tacoma accident" (1940), and other unique effect, which occurred on May 2010 at the Volgograd road bridge designed as continuous beam;
- with a single wind pulsation, related to the shape and rigidity of the object, the surrounding topography and spatial planning, etc.;
- with irregular flows of inertial and partly spring-mounted transport and etc.;
- with service transmitted by the ground (and on other transfer mechanisms) seismic-waves of natural and man-made origin,

[*] *V.V.Kulyabko* Pridneprovskaya State Academy of Civil Engineering and Architecture, str. Dnepropetrovsk, Ukraine

including the interaction of the structures near water bodies with such of their consequences as tsunami.

These and many other types of interaction of structures with other subsystems and their combinations give rise to the new versions of the dynamic loads. These actions in quantitative and qualitative terms may vary, which makes it difficult to otherwise attempt making "simplified" regulation, in addition to direct simulation of dynamic interaction of objects with the actual parameters in the time domain with account of all types of nonlinearities.

Sec.4.2. Classification of Dynamic Loads

Let's consider some classification of dynamic loads and actions, which can under certain conditions cause the accident, the destruction of buildings and structures (and also at lower levels of actions discomfort people in buildings: pain, a decline in productivity, etc., [4.6] and some of the requirements of the international standards on permissible vibration in various buildings and structures of the type of the offshore platforms, for example, etc.):

1. Monogarmonic force effect

This is a typical action on building structures (denote them as "receiver vibration N") from a technological equipment ("source S"), which has a rotating circular frequency Θ and eccentricity e of the mass m_{vr}. The amplitude of the perturbation force is equal to $P_o = m_{vr}e\Theta^2$. This, the most famous of the history of civil engineering configuration corresponds to the work of electric machines, centrifuges, fans, pumps, vibro-drivers and vibrating tampers, vibrorollers, mechanical and electromagnetic vibrators in the vibroplatforms, screens, etc. The hazard from the performance of such kind of equipments (up to the emergency destruction) arises, first of all, at resonance phenomena (the coincidence of the frequency of forced vibrations of the machine with one of the frequencies of oscillation of the supporting floors in building or bridge or structures at an elastic foundation).

2. Monogarmonic kinematic action

If the source S is located without the direct transfer of the force P_0 on the receiver N, it may transfer some harmonic kinematic movement (with the delay or without it) supporting or other points of the N through the "mechanism of the transmission of vibrations M". The hazard of such a dynamic load is similar to the previous view. Some parts of seismograms with single-frequency type present an example of such action. This version associated with the seismically dangerous areas, can be considered in accordance with losses as the most destructive action on structures. In addition, there are numerous kinds of technological, transport, urban, etc. of the seismic data.

3. Polyharmonic action

This type of dynamic loads has its dangerous side. For example, the use of friction damping devices or connections in the conditions of a two-frequency vibrations of structures may be useless as the dry friction forces (under certain modes, and the ratio of the amplitudes of velocities of the high and low frequencies) can decrease the order, which may lead to accidental condition of the object. In the experiment on testing of the effectiveness of the work of the joint on high-strength bolts coefficient of friction "steel-steel" was obtained: prior to the application of polyharmonic action equal to 0.30, and at the time of his application – 0.03. Such actions, for example, on a sandy clay base, also lead to a phenomenon known in soil science as "the dilution of the sand. "

4. Random dynamic action

This is the most common and, of course, the most prevailing process of dynamic actions. Herewith there is always a certain background, registered be vibration-survey equipment in the form of microseisms. However er these dynamic actions in the form and on the complexity of certain experimental data are one of the most difficult. These are almost all natural causes (seismic, wind, waves) and many anthropogenic one (technological, transport, etc). The most correct but difficult way is to collect and process the necessary amount of implementations. However, the further application of the results of such inputs actions is difficult even for linear problems using the FEM in spatial models of structures. For nonlinear models and systems of the type of damping devices a probabilistic approach is possible without application of linear algebra and various methods of linearization, for example, through the "direct" research random responses reactions subsystems in the time domain. The hazard for the buildings from the random actions may appear not so much on the type of process, as of his level. The exception noted in clause 3 by the phenomena of the kind of vibration smoothing of dry friction forces, etc.

5. Pulse single action

Pulse single action is the single wind gusts and storms, tornadoes; peak shock and moving in case of earthquakes and tsunamis, etc. This is power pulses. And kinematic pulses can be called the perturbation of the flow of traffic on a single unevenness of the road, for example, on the bridge, and impulses that are transmitted on the structures through the soil or other body (the transmission mechanism) - pile-driving, methods of soil compaction by rammers, industrial explosions, etc. The hazard for buildings is, firstly, in large amplitude of the power or the kinematic impulse, and secondly, the proximity of the pulse duration to the period of oscillations of the object at any of natural mode.

6. Periodically repeated pulse actions

These dynamic loads are characterizing by a certain period of power or kinematic pulses, described in clause 5. Performance of the various stamping machines, hammers, pile-driving pile drivers is considering. In this case, the perturbation of the flow of traffic on uneven road bridge repeated depending on the length of the links of the railway track ("traces" of joints remain even after welding parts in the scourge of non joints way) or on the condition of the road. Similar can be the impact on the near and distant buildings from any rail transport, including tram, subway, etc. The hazard of such actions associated the phenomenon of resonance described in clause 1 when the return period of the pulse will be close to the period of natural oscillations of structures.

7. Natural seismic actions

The practice of construction all over the world shows that this type of dynamic loads and impacts is the most destructive for humanity. This event in some measure includes all of the above six types of actions. There are various theories of design and the scale on the relative estimates of the levels, causes and consequences of earthquakes. We single out two basic methods of structural design on the dynamic load: first, the spectral method in which according to the place of location of the construction site and to the foundation of the buildings on that site seismic magnitude is determined; the second method is to direct dynamic design of accelerogram recorded in the area of the construction site.

8. Seismic activity: industrial, transport, urban

A demonstrative example of this type of dynamic loads was the fluctuations of the whole microdistrict (N), remote from sources S on 1-2 km. The author studied the mechanisms of transfer of oscillations M from the road and railway transport, from the industrial ground noise and from dozens of powerful compressors with numbers 125 and 150 rpm (crank-and-rod mechanism of piston compressor "betrayed" the amplitude of the perturbation of the horizontal forces of up to 100 tnf on the environment through the sole of the foundation of the size of the order of 12×18 m). It turned out that near the unit of these latter sources on the ground move was 600 mkm, the blind area the studied buildings 5-10 mkm (more than 1 km from the plant), and because of the resonance with one of his own forms of 9-storeyed hospital building (the rotation of the plan) covering had to 200 mkm! (Other sources in this example, passed the vibration on the order less). It is important to note that to the destruction of the hospital building this prom-seismic activity did not lead, but create a very harmful effect on the operations and medical treatment.

9. Mobile loads

Mobile loads are varied and yield little to engineering correct analysis for the last more than 160 years (railway transport, motor transport, the flow of air wave).

These should include the entire rail and road transport, air currents, explosive shock waves, wind waves of the seas and oceans, tsunami, etc. Unfortunately, there are very few methods and suggestions for the correct considering high-speed interaction of inertial media flows and assets with the inertial structures - with the account of random intervals, delay, etc. Are in need of improvement models of dynamic interaction between the buildings and platforms structures (remember the collapse of an overpass in the earthquake in Kobe, Japan) with movable bridge and the tower (attached) cranes.

10. Progressive collapse

The progressive destruction of the building structures and facilities can be combined to reduce classification with local dynamic actions of anthropogenic type and blows on structures, buildings, grounds. Note that this group is one of the most poorly known, but very dangerous and complex upon mechanics of impacts.

This theme, of course, is indicated by an extremely complex. But this does not mean that it should be avoided. The need to study and discuss as often as possible in order not to repeat the failures of the type listed in the work [1], to reduce the damage from the typical "dynamic" disasters.

References

*Ch.*1

1.1. DmitrievF.D. (1953) The collapse of engineering structures.Stroyizdat Publ.House, M., 186p. (Крушения инженерных сооружений).

1.2. Mac Keig, X.Tomas (1967) Accidents of Structures, Stroyizdat Publ.House, M., 148p. (Строительные аварии).

1.3. KikinA.I., VasilevA.A., KoshutinB.N.and others (1984), Increase of durability of metal structures of industrial buildings, Stroyizdat Publ.House, M.,301p. (Повышение долговечности металлических конструкций промышленных зданий)

1.4. Mizymskii I.A. (1965), The collapse of steel structures for the faults of the design, Proc.XXIII Conf.LICI, Engineering Structures (Аварии стальных конструкций из-за недостатков проектирования)

1.5. Lashchenko M.N. (1969), The accidents of metal structures of buildings, Stroyizdat Publ.House, L., 184p. (Аварии металлических конструкций зданий и сооружений).

1.6. Federal Low №2446-FZ (03.05.1992), On Safety (О безопасности).

1.7. Federal Low №384-FZ (12.30.2009), Technical regulation on buildings safety (Технический регламент о безопасности зданий и сооружений.

1.8. GOST 27751-88 The reliability of building structures and foundations. The main provisions on the design. (Надежность строительных конструкций и оснований. Основные положения по расчету).

1.9. Rzanitsyn A.R. (1978), Theory of Structural Reliability in Design, Stroyizdat Publ.House, M., 237p. (Теория расчета строительных конструкций на надежность).

1.10. GOST.51901.1-2003, The management of risk. Risk analysis of technological systems. (Менеджмент риска. Анализ риска технологических систем).

1.11.GOST R 51901.14-2005. The management of risk. Method of the structural scheme of reliability. (Менеджмент риска. Метод структурной схемы надежности).

1.12. Bekyaev B.I., Kornienko (1968), The causes of accidents steel constructions and methods of their elimination, Stroyizdat Publ.House, M., 206p. (Причины аварий стальных конструкций и способы их устранения).

1.13. Augustin Ya., Shlelzevskii E. (1978), The collapse of steel structures, Stroyizdat Publ.House, M., 177p. (Аварии стальных конструкций).

1.14. Shkinev A.N. (1984), The collapse of steel structures, Stroyizdat Publ.House, M., 320p. (Аварии в строительстве).

1.15. Message, Gosstroy RF (04.05.1999), №BE-1080/19 On measures to prevent accidents on construction and operated buildings and structures. (О мерах по предотвращению аварий на строящихся и эксплуатируемых зданиях и сооружениях).

1.16. Report (2004), Accident of buildings and structures on the territory of the Russian Federation in 2003" The Russian public Fund "For the quality of construction". M. (Аварии зданий и сооружений на территории Российской Федерации в 2003 году)

1.17. Melnikov N.P. (edit.) (1980), Metal structures: a Handbook for the designer, Stroyizdat Publ.House, M., 776p. (Металлические конструкции: Справочник проектировщика)..

1.18. Streletckyi N.S. (edit.), (1952), Steel Structures, Stroyizdat Publ.House, M., 852p. (Стальные конструкции).

1.19.Val V.N.,Gorochov E.V.,Uvarov B.Y.(1987), Reinforcing steel frames one-storey industrial buildings with their reconstruction, Stroyizdat Publ.House, M.,220p.(Усиление стальных каркасов одноэтажных производственных зданий при их реконструкции).

1. 20.www.pamag.ru / Portal "Science and safety".(Портал "Наука и безопасность").

1.21. Machutov N.A. (2009), Methods and safety standards strategically important buildings and structures / Abstract / / IV international conference "Prevention of accidents of buildings and structures" M. (Методы и нормы обеспечения безопасности стратегически важных зданий и сооружений).

Ch.2

2.1. MNIITEP (2006), Recommendations for the protection of high-rise buildings from the progressive collapse, M. (Рекомендации по защите высотных зданий от прогрессирующего обрушения).

2.2. MDC 20-2.(2008), Temporary recommendations on the safety of long-span structures from the progressive collapse in case of progressive collapse due to accidental actions.(Временные рекомендации по обеспечению безопасности большепролетных сооружений от прогрессирующего обрушения при аварийных воздействиях).

2.3. FL №123, (08.22.2008), Technical regulation on fire safety requirements, (Технический регламент о требованияхи пожарной безопасности).

2.4. SP 2.13130 (2009), System of fire-prevention protection. Provision of fire protection objects, (Системы противопожарной защиты. Обеспечение огнестойкости объектов защиты).

2.5. STO 01422789-001(2009), Design of high-rise buildings, M. TSNIIEP dwellings. (Проектирование высотных зданий).

2.6. Roitman V.M.(2009), The background of fire safety of high-rise buildings, M.,ISA MGSU, 107p.

2.7. Roitman V.M.(2001), Engineering solutions on the assessment of the fire resistance of the designed and reconstructed buildings, M. Pozhnauka, Publ. House, 381p. (Инженерные решения по оценке огнестойкости проектируемых и реконструированных зданий).

2.8. Agafonova V.V., Rodionov G.A., Roitman V.M. (2010), Hazardous effects affecting the resistance of the buildings at the combined accidental actions with the participation of fire. In the book: Construction, formation of the living environment, Reports at 13 Conf. Junior Sci., M. MGSU.(Опасный эффект влияющий на устойчивость зданий при комбинированных особых воздействиях с участием пожара).

2.9. Telichenko V.I., Roitman V.M. (2010), Ensuring the resistance of buildings in case of combined accidental actions with participation of fire present the basic element of the complex security system. Collection of scientific works: The prevention of accidents buildings, 9, M. 15-29pp. (Обеспечение стойкости зданий при комбинированных особых воздействиях).

Ch.3

3.1. Freund R., Wilson W., Sa P. (2006), *Regression Analysis*, Elsevier Science, 480pp.

3.2. Cramer D. (2003), *Advanced Quantities Data Analysis*, Open Univ. Press, 376pp.

3.3. Gross J.L. (2005), *Graph Theory and its Applications*, Wesley & Sons, 800pp.

3.4. Aitkin C.G.G., Taroni F. (2004), *Statistics and the Evaluation of Evidence for Forensic Scientists (Statistic in Practice)*, J.Weley & Sons, 540pp.

3.5. Bedford T., Cooke R. (2001), *Probabilistic Risk Analysis: Foundations and Methods,* Cambridge Univ. Press, 414pp.

3.6. Calafiore G., Dabbene F.-Editors (2006), *Probabilistic and Randomized Methods for Design under Uncertainty*, Springer, 457pp.

3.7. Klir G.J., Bo Yuan (1995), *Fuzzy Sets and Fuzzy Logic: Theory and Applications*, Prentice Hall, 592pp.

3.8. Gumbel E.J. (1967), *Statistics of Extremes,* Columbia University Press, New York.

3.9. Lew H.S. (2005), *Best Practice Guidelines for Mitigation of Building Progressive Collapse*, Building and Fire Research Laboratory, National Institute of Standards and Technology, Gaithersburg, Maryland, USA, 20899-8611.

3.10. Raizer V. (2009), Reliability of Structures. Analysis and Applications, Backbone Publ. Co., USA145p.

3.11. Mkptychev O. V. (2000), *Reliability of Multiple-unit Bar's Systems of Engineering Structures*, Manuscript of doctorial thesis, Moscow (in Russian), 493p.

3.12. Raizer V.D., Mkrtychev O.V. *"Nonlinear Probabilistic Analysis for Multiple-unit Systems"* Proc. 8th ASCE Specialty Conference on Probabilistic Mechanics. July 2000,Univ. Notre-Dam. IN

Ch.4

4.1. Eremin K.I. (2010), Chronicle of accidents buildings and structures that have occurred in 2009. / / The prevention of accidents of buildings and constructions: Collection of scientific works, 9. - M.: RAACS, MGSU, WELD 5-15p. (Хроника аварий зданий и сооружений, произошедших в 2009г. // Предотвращение аварий зданий и сооружений).

4.2. Kulyabko V.V. (2010), Dynamics - and the cause of the accidents with structures, and the way of their prevention / / the Prevention of accidents of buildings and structures: Collection of scientific works, 9. - M., RAACS,

WELD. 86-90p. (Динамика – и причина аварий сооружений, и путь их предупреждения // Предотвращение аварий зданий и сооружений).

4.3. Korenev B.G., Rabinovich I.M. (2 ed.), (1984), Dynamic design of buildings and structures: Handbook for design, M., Stroyizdat Publ.House, 303 p. (Динамический расчет зданий и сооружений: Справочник проектировщика).

4.4. KorenevB.G. Rabinovich I.M. (ed.), (1981), Dynamic analysis of structures on the special actions: Handbook for design, M., Stroyizdat Publ. House,215 p.(Динамический расчет сооружений на специальные воздействия: Справочник проектировщика).

4.5. KorenevB.G. Smirnov A.F. (ed.),(1986), Dynamic design of special engineering structures: Handbook for design, M.: Stroyizdat Publ.House 461p.(Динамический расчет специальных инженерных сооружений и конструций: Справочник проектировщика).

4.6. Kazakevych M.I., Kulyabko V.V. (1996), Introduction to vibroecolodgy of buildings and structures, Dnepropetrovsk: PSACA, 200 p. (Введение в виброэкологию зданий и сооружений).

4.7. The Ministry of Construction, Ukraine (2006), Construction works in Ukraine seismic regions, 84p. (Строительство в сейсмических районах Украины).

Part II. Real Bearing Capacity of Structures

Ch.5. Buildings and Structures as Complex Natural and Man-made System

Grunin I.Y., Budko V.B., Aleksandrov S.V.[*], Morozov P.A.[†], Merkulov A.F.[‡]

Sec.5.1. General Comments

The text introduced in this chapter is the result of many years of research conducted by the Institute of Technology "VEMO" and it's partners in the Consortium of engineering-technical audit on the study of the influence of natural conditions on the safety of complex natural-man-made systems

It should be noted that this research being of a strictly applied nature, has been carried out by experts and specialists of non-destructive control for the decision of tasks on specific objects.

At the first stage of the research a theoretical model the scientific concept of the natural-industrial systems (NIS) was adopted. This concept was developed by Russian scientists in the past 20 years on the basis of geological-geographical achievements; mainly engineering geology, as well as landscape architecture, geocryology, geomorphology and hydrogeology, combined with the experience of survey, design, construction and operation of engineering structures.

At the second stage, created on the basis of the results of studies at the first stage, the theoretical model was called "ecological and industrial systems" (EIS) and has included the concepts of the NIS and active factors of anthropogenic action on the system. It was established in the course that a number of signs indicating the presence of processes in the EIS, with the source that generates these processes outside the EIS.

It was decided to research ecological factor– "Ecological system" (ES) separately, since it affects environment, while in the meantime when it comes to the immediate object of research, it should be reverted to the conventional meaning of NIS, modifying it in accordance with the increasing complexity of problems solved by the addition of the word «complex». In the final version the term, «the Complex of natural-industrial system» (CNIS) was adopted.

---

[*] SLL «Institute of Technology energy surveys, diagnostics and non-destructive testing "VEMO"», Moscow.
[†] Institute of terrestrial magnetism, ionosphere and radio wave propagation them. N .V. Pushov, Russian Academy of Sciences (IZMIRAN), Moscow. Troitsk, Moscow region.
[‡] SLL «Russian scientific-research Institute of special methods of survey».

At the third stage the area of studies was significantly expanded and deepened. For understanding of the processes taking place in CNIS, an extensive archival research has been carried out with the aid of written sources, cartographic materials, etc.

The studied processes are common in the study of the totality of conditions and interactions of components of the natural environment and engineering facilities at all stages of the life cycle - from pre-design up to dismantling and recycling, taking into account the changing of technical actions.

Thus, the subject of studies of the formation and investigation of the model CNIS has a set of building structures and engineering systems of a certain ecological and industrial complex It provides regular functioning of the EIS in the conditions of dynamic equilibrium of all forming components and specific components of the natural environment. The influence of mentioned above on engineering structures may cause the accidents with possible severe social and environmental impacts, and not in isolation, but in the relationship.

The effect of CNIS is considered in the framework of the ES. Thus, buildings and structures in the conditions of site development, such as industrial site or route of linear structures, one can see as CNIS, exerting influence on the ES.

Sec.5.2. Accident scenarios

It should be noted that the prediction of the development for accident scenarios for the designed objects means a clear awareness of such opportunities. It is necessary to mention, that the scenario of the accident can be developed for any object, and not only for one. In fact, it is a kind of solution to the inverse problem, which in many ways is ideal. A number of scenarios of possible development of the accident for any object is in fact, infinite, as any linear shame describing the scenario, bypassing all bifurcation points in the giant tree of possible events and gives only one definite version of the development of these events. It should be also noted that at the beginning of the construction of the most predictive schemes are undergoing already a radical change.

Sufficient degree of trustworthiness of accident scenarios acquire at the consideration of operated objects with an expert assessment of their technical condition. The most positive forecast scenarios are developed for the objects that already have clear signs of destructive processes. The forecasting process has enough decision procedure and is workable at this stage. The experience accumulated by experts in the field of construction and industrial safety in the preparation of such forecasts is huge. A considerable part of the knowledge and experience by experts is based on the study of scenarios and mechanisms of actual accidents.

The material presented here, aimed at the empowerment of experts when developing scenarios alleged accidents by taking into account additional factors, not always fall within the field of view of experts.

Sec.5.3. Problems Arising in the Development of Accident Scenarios

Experts, developing emergency scenarios, are criticized often enough for the categorical wording, the vagueness of the wording or inaccuracies temporary forecasts.

This kind of accusations is actually groundless. Expert intuitively chooses the worst-case scenario because he can imagine a model of the process of destruction of the object. The conventional matrix «cleaned» from a number of stabilizing factors, often putting off an accident at a significantly removed from the forecast time. Thus it is necessary to emphasize, that in most cases the forecasts are accurate and reflect the possibility of an emergency scenario.

This should take into account the following factors:

– Forecast scenario for the development of the accident allow from the one hand to take steps to eliminate the causes of emergency processes at an early stage, and from the other hand - for the conduct of such activities receive the amount of distortion, which requires a full review.

– The forecast scenario for the development of the accident focuses most often on the assumption of further development of destructive processes that are already embedded in the object and may not provide the options of the impact of random provoking factors, which can accelerate or otherwise change the course of the process.

– In the course of checking the effectiveness of implemented forecasts there was detected a really effective natural factor, which slows down the processes of development of the accidents, which resulted from a mechanical destruction of building structures under the action of hazardous exogenous processes. Factor is related to the so-called recreational properties of the soil floor of the object's location. The study of this factor is taken as a promising development.

– Factors of influence of a planetary scale: atmospheric, geological and hydrological should be evaluated not only at the design stage, but also at the stage of survey, taking into account their natural variability in time and changes occurred in CNIS for the period of operation.

All kinds of accidents of buildings may be conditionally divided into four groups:

1. The collapse of building connected with deformation processes.
2. Explosions.
3. The fires.
4. The emissions.

The main kinds of the collapse of building structures:

1. Collapsed structures under its own weight at the stages of construction and the start of operation due to the violations, committed during design and construction.

2. The collapse of structures, caused by deformations under effects of external factors.

3. Spontaneous collapse of structures worked out their service life, due to excessive wear of structural materials.

Let us consider the main promising development in the framework of the research CNIS, associated mainly with the study of the deformation processes, leading to accidents, and associated partly with the explosion of the natural and industrial character.

The basic developments, shown below, related to the development of the express methods of diagnosis of dangerous processes and aimed at development of technologies for the detection of an emergency processes in the early stages of their development. It is connected, first of all, with the necessity of taking urgent operational decisions on the prevention of accidental collapse of structures, the risk of which can be found already in the course of primary surveys.

Sec.5.4. Development of Visual and Measuring Control Technologies

One of the promising directions is the development of technologies of visual and measuring control in the construction expertise, allowing in the course of primary examinations and express-diagnostics to identify defects structures, evaluate their character, the degree of danger and the possible consequences of development.

Consider this technology at the example of rock and reinforced masonry structures.

The main types of damage and defects of rock and reinforced masonry structures, characterizing the place of their location, the possible reasons and possible consequences of the development are given in Table. 5.1.

Table 5.1

Typical damages and defects in the masonry

Types of damage and defects, their location and characteristics of the detection	Probable reasons of origin	Possible consequences
The curvature of horizontal and vertical lines.	Non-uniform deformations of soil grounds. It is possible the appearance of characteristic cracking.	Decrease of bearing capacity, the development of cracks.
Walls buckling	Lateral pressure of the soil, various materials, placed in bulk over the walls, the effect of horizontal reactions sprung structures. Increase (against design) eccentricities of vertical loads. The greater flexibility of the wall in height because of the gap or lack of intermediate ties. The displacement on the basic beams trials, slabs or floors to the edge of the wall. The transfer of unacceptable force action on the clutch, not accumulating the adequate strength. Unilateral thawing of masonry, performed by the method of freezing. Thermal deformation.	Reduction of the bearing capacity of the wall; the emergence and development of cracks; stratification of masonry; the destruction of the materials masonry; loss of stone fragments, further damage to the masonry. The possibility of a collapse of a section of the wall.
The deviation of the walls or their individual plots from the vertical	Non-uniform deformations of soil; the lack of cross-ties, or their break-up Lack of cross-ties, or their break-up	Forthcoming and development of cracks in the masonry, reduction of the carrying capacity. With the development of the sedimentary processes - the ability to breakaway and collapse sections of facade masonry
Spalls groups of the corners, gaps, potholes, ruts, etc.	Defects of construction, mechanical effects in the process of operation (blows of vehicles, punching of holes and slots for various purposes, etc.)	The possible decrease of bearing capacity

Types of damage and defects, their location and characteristics of the detection	Probable reasons of origin	Possible consequences
In the field of damage to the outer layer (plaster, cladding, etc.)	The accumulation of moisture from atmospheric precipitation on the damaged areas of the outer surfaces of the walls and its capillary absorption by materials of masonry in the thickness of the wall	The development of destructive processes with subsequent micro- and macro destruction of rock and mortar
The influence of the open placement of equipment, producing steam and moisture	The condensation of moisture on the surface of the walls, entry of splashes	The development of destructive processes in the masonry with the subsequent progressive collapse
In parallel or cornice part of the exterior walls under the windows, niches, in the area of location of the drain-pipes	Damage to the roof in the cornice zone, poor execution of the landfall of the waterproofing carpet to the wall. <hr> Damage to gutters, lack of drip edges, damage discharges, funnels and drainpipes <hr> Insufficient or reverse bias, lack of takeaway cornice eaves	The development of destructive processes in the masonry with the subsequent progressive collapse
Above the windows, gates, doors, local exhaust ventilation openings with the possibility of creation in the winter frost and ice.	Exfiltration of air from the premises of the building with condensation of moisture	The development of destructive processes in the structures of lintels. Development of deflection with the destruction of the elements
In the basement of the walls	Damage, poor performance or lack of waterproofing; the low waterproofing position in relation to perimeter walks, damage of perimeter walks or sidewalk	The development of destructive processes in the masonry, caused by the alternate freezing and defrosting of humid areas. The destruction of the masonry. Reduction of the bearing capacity of the wall

Types of damage and defects, their location and characteristics of the detection	Probable reasons of origin	Possible consequences
Wetting of the inner surface of the walls in the whole space or in different zones	The mismatch of actual temperature and humidity in the room adopted in the design (lack of ventilation, the changes of the technological process)	Reduction of the strength parameters of masonry
	The mismatch of actual thermo-physical characteristics of materials adopted in the design, lack of thermal insulation of separate zones	
The humidification of the walls in the zones of sanitary equipment, pipelines, containers with liquid	Equipment failure, leaks from pipelines and tanks	Reduction of the strength characteristics of masonry with the development of destructive processes
	Constant condensation on the surface of pipelines, containers with liquids and etc.	
Salty appearance on the outer or inner surface of the walls	The transfer of salts, included in the materials of the wall, on the surface with their high-dose supplements (additive and mortar)	Reduction of the strength characteristics of masonry with the development of destructive processes
Peeling, cracking or delaminating of paint and varnish coatings	The deformation and the destruction of the walls material under the paint-and-lacquer coating	Reduction of the strength characteristics of masonry with the development of destructive processes
	Deformation of alternately freezing and defrost moisture	
	The discrepancy of paint and coatings to thermal and moisture regime of the air or of a chemical aggressiveness of the operational environment	There is no effect to carrying capacity of the masonry in the case of timely elimination
	Violation of the rules of the paint devices	
Peeling plaster coatings, or textured layers with falling out the separate pieces	Deformation or destruction of wall materials under the plaster layers	Reduction of the strength characteristics of masonry
	The difference in shrinkage or temperature deformations of the plaster layer and the wall	
	The penetration of moisture under the plaster layer with subsequent repeated cycles of freezing-thawing or wetting-drying	

Types of damage and defects, their location and characteristics of the detection	Probable reasons of origin	Possible consequences
	High-temperature heating (technological or in the event of fire)	There is no effect to carrying capacity of the masonry in the case of timely elimination
	Defects in workmanship or in applying of coatings	
The loose structure of the plaster layer	The alternate freezing-thawing of material plaster layer in the moist condition	There is no noticeable effect to carrying capacity of the masonry in the case of timely elimination
	Wedged action of moisture at alternating wetting-drying	
	Dissolution or washing away of the components of the material with water	
	Chemical impact on the materials of the plaster layer	
Cracks in the masonry, have the character of a parabolic curves, the branches of which differ down on both sides of the middle part of the building	Deformation of ground in the middle part of the building	Reduction of the bearing capacity of the walls in the zone of the location of cracks, decreasing of the spatial rigidity of the building
Cracks, the disclosure of which is increased to the top of the sloped or have the character of a parabolic curves, diverging to the bottom on the edges of the building	Deformation of the ground at the end parts or the presence of a rigid inclusion under the central part of the building	Reduction of the bearing capacity of the walls in the zone of the location of cracks, decreasing of the spatial rigidity of the building
Cracks, close to the vertical, the disclosure of which is increased to the top	Rift of the building, due to the rigid support in the ground under the crack	Reduction of the bearing capacity of the walls in the zone of the location of cracks, decreasing of the spatial rigidity of the Building
Close to a vertical crack to the same disclosure in height with an offset of the vertical part of the building to one side of the cracks on the other	Deformation of the ground under the part of the building	Reduction of the bearing capacity of the walls in the zone of the location of cracks, decreasing of the spatial rigidity of the building

Types of damage and defects, their location and characteristics of the detection	Probable reasons of origin	Possible consequences
V-shaped cracks on line of addition of a new building to the previously existing or in place of difference of heights of a building	Different degree of soil compaction or different pressure on the ground on both sides of the line of additional building or difference of heights	Reduction of the bearing capacity of the walls in the zone of the location of cracks, decreasing of the spatial rigidity of the building
Vertical cracks with the opening of the 0.1-0.5 mm, crossing two or more rows of masonry, and the number of cracks two and more per 1 m due to vertical loading wall, and bundle of masonry	A significant overload of masonry, reduced strength of the materials used in the structure and respectively reduction of the strength characteristics of masonry	Reduction of the strength characteristics of masonry with the development of destructive processes
Horizontal and out-of-square cracks in open joints of a laying of the ordinary, wedge-shaped or arched lintels; vertical cracks in the middle of the span, it is also possible to fall out of some individual stones	Overloading of bricklaying, reduced strength of materials, lack of reinforcement, non-uniform deformations of grounds	Reduction of the strength characteristics of masonry with the development of destructive processes
Horizontal cracks in seams on a laying of walls, exposed to the horizontal loads, and it is possible with a shift in horizontal seams or stepped inclined toothing	Overload of masonry, reduced strength of materials, lack of reinforcement, non-uniform deformations of grounds	Reduction of the carrying capacity
Small cracks of masonry materials below the supports and supporting parts of the beams, trusses, lintels differ from the location of the load	Overload of masonry, as well as the insufficient depth of the supporting part, the absence or insufficient carrying capacity of the supporting cushion	Decrease the strength of the masonry to the state of accident
Vertical and inclined cracks in the upper part of the building, in the field coupling of longitudinal and transverse walls which are loading in a different way	Various deformability of the loading of walls in a different way due to different stresses in the masonry and appearance the creep under the long-term load	Reduction of the bearing capacity of the walls in the zone of cracks, decreasing of the spatial rigidity of the building

Types of damage and defects, their location and characteristics of the detection	Probable reasons of origin	Possible consequences
Vertical cracks in the upper part of the pilasters, being supports to beams and trusses, in the field of coupling of pilasters and masonry walls	Various deformability of the loading of walls in a different way due to different stresses in the masonry and appearance the creep under the long-term load and also horizontal forces arising in trusses and beams with oscillations of temperature, setting of the foundations	Reduction of the bearing capacity. Decreasing of the spatial rigidity of the building
Cracks of the V-shaped form in the upper part of the building	Various deformability of the loading of walls in a different way due to different stresses in the masonry and appearance the creep under the long-term load and also horizontal forces arising in trusses and beams with oscillations of temperature, setting of the foundations and thrust due to disorders of the rafter system covering of the building	Reduction of the bearing capacity. Decreasing of the spatial rigidity of the building
Vertical cracks with the opening of the 0.1-0.3 mm in the masonry of the longitudinal walls at the lower floor, on the ends of the lintels, beams, plates, reinforced belts, due to breaking of the longitudinal walls from the end and transverse one	Longitudinal temperature-humidity deformations of walls or floors in the change in the average temperature of section	Reduction of the strength characteristics of masonry with the development of destructive processes
Cracks with the opening of up to 10 mm and more, the gap in the masonry of the middle part of the building at all its height	The absence of reinforcement in the sedimentary seams or lack of reinforced belts for the perception of temperature-humidity deformations	Decrease of the strength of the masonry in the zone of the cracks
The slanting cracks in the nodes of extreme apertures of the first floor	The absence of the temperature-sedimentary seams or lack of reinforced belts for the perception of temperature-humidity deformations	Decrease of the strength of the masonry in the zone of the cracks

Types of damage and defects, their location and characteristics of the detection	Probable reasons of origin	Possible consequences
Peeling of the surface, weathering of the outer layers, increased porosity, lower density of loose structure, chalking, loss of individual parts of the material	Influence of chemically aggressive operational environments	Reduction of the strength characteristics of masonry with the development of destructive processes
	High-temperature heating of the technological sources or fire effects in the event of fire; moisturizing	
	Biochemical effects of microorganisms, fungi, mosses, etc	
	Biochemical effects of trees and shrubs	
	The alternate freezing-thawing in humid condition in case of insufficient frost resistance, alternate wetting-drying	

Sec.5.5. Hazard of Dangerous Natural Processes

The second development relates to the necessity for express evaluation of the degree of hazard of dangerous natural processes (DNP), influencing on the object.

The technology of obtaining data on this issue involves the evaluation of the DNP on the cards of the natural hazards, developed practically in all regions, updating data on the results of visual inspection and route studies and identification of the external manifestations of this DNP to the object under study.

Let us examine this question on the example of the results obtained in the study of the subject in the Urals Federal district (Table 5.2), taking into account criteria DNP (Table 5.3).

Table 5.2

An analysis of the manifestations of the actions of dangerous natural phenomena on the building structures

Name of danger	Updated forecast	The existence of manifestations
Geological and geophysical hazards		
Seismic danger	3 – dangerous, 6-7 grades 4 – moderately dangerous, 5-6 grades	Not found

Name of danger	Updated forecast	The existence of manifestations
Category of karst hazard	5 – slightly dangerous 6 - the territory of the possible manifestation of karst in techno-genic actions	Not found
Category of karst-suffosion danger	1 – very dangerous 2 – dangerous	Watering of the zones around foundation. The destruction of blind area. Soil deforma-tions
Danger of landslide phenomena	3 – dangerous 4 – moderately dangerous	Not found
Danger of mudflows	4 – moderately dangerous 5 – slightly dangerous	Not found
Resistance to dynamic loads (vibration)	1 – unstable 2 – relatively stable	Cracks sedimenta-ry nature. The zone of stratification
The hydro-geological and geocryological hazards		
Geocryological danger	5 - areas with dangerous manife-stations of the processes of less than 1% and moderately danger-ous - less than 10% of the area	The destruction of blind area, areas of subsidence and buckling of soils
The danger of frost heaving soils	2 – dangerous 3 – moderately dangerous	The destruction of blind area, areas of subsidence and buckling of soils
The danger of change of a level of subsoil waters	2 – average	Not found
The degree of under-flooding of under-ground waters	2 – potentially underflooding 3-non- underflooding	The flooding of the zones around foundation by upgrade water
The degree of danger of development of plane and gully ero-sion	2 – moderately dangerous	Not found
The ground water, which are characte-rized by different types of aggression against concrete	1 – general acid 2 – carbon-dioxide 3 – sulfate	The destruction of the material of brickwork

Name of danger	Updated forecast	The existence of manifestations
The index of the non-equilibrium ground-water in relation to the calcium carbonate	3 – < – 1 2 – from 0 to – 1	The destruction of the material of brickwork
Atmospheric hazards		
The danger of heavy rains in points	7 points	The flooding of the zones around foundation, lea-kage
The danger of the strong snowfall in points	4 points	Not found
Risk of thunderstorm and the hail	4 points	Not found
The danger of strong winds (hurricanes, typhoons, tornadoes, squalls)	2-average	Not found
Danger of extremely low temperatures	2-average	Stratification and buckling of a lay-ing. The destruc-tion of the material of the lentils
Danger of extremely high temperatures	1 – high 2 – average 3 – low	Cracking of roof-ing material
Anthropogenic hazards associated with the depletion and the loss of the elements of the natural-resource potential		
The danger of viola-tions of the regime of the watercourse	1 – high 2 – average	Watering of the zones around foundation
The danger of viola-tions of the hydro-chemical regime of the ground water	1 – high	The dissolution of the material maso-nry mortar
The danger of activa-tion of karst processes (caused anthropogeni-cally)	2 – average	Not found

Name of danger	Updated forecast	The existence of manifestations
The danger of activation of karst-piping processes caused (anthropogenically)	1 – high	Watering of the zones around foundation The destruction of lentils. Subsidence of soil

Table 5.3

The categories and criteria of dangerous natural processes

Indicators used in the assessment of the degree of danger of the natural process (DNP)	Hazard category processes	Indicators
Landslides		
Areal affection of territories,%	extremely hazardous (catastrophic)	more than 30
	very dangerous	11–30
	Dangerous	1-10
	moderately dangerous	0.1–1
The square of a one-off appearance on one area, km^2	extremely hazardous (catastrophic)	1–2
	very dangerous	1–0.5
	Dangerous	0.01 –0.5
	moderately dangerous	less than 0.01
The volume of the captured rocks during the one-time manifestation, million m^3	extremely hazardous (catastrophic)	10 – 20
	very dangerous	5 – 10
	dangerous	0.001 – 5
	moderately dangerous	till 0.001
Speed offset	extremely hazardous (catastrophic)	till 5m/sec.
	very dangerous	till 2m/sec.
	dangerous	1 – 2m/sec. (1 – 10m/day)
	moderately dangerous	1 – 5m/day (5 – 10m/month)
Frequency of occurrence, units per year	extremely hazardous (catastrophic)	0.01 – 0.1
	very dangerous	0.03 – 0.1
	dangerous	0.1 – 0.2
	moderately dangerous	till 1

Indicators used in the assessment of the degree of danger of the natural process (DNP)	Hazard category processes	Indicators
Sill		
Areal affection of territories,%	extremely hazardous (catastrophic)	more than 50
	very dangerous	10 – 50
	dangerous	5 – 10
	moderately dangerous	less than 5
The square of appearance on one area, km^2	extremely hazardous (catastrophic)	till 5
	very dangerous	till 3
	Dangerous	till 1
	moderately dangerous	less than1
The volume of simultaneous removal, million m^3	extremely hazardous (catastrophic)	till 5 – 10
	very dangerous	till 1 – 3
	Dangerous	till 0.5 – 1
	moderately dangerous	0.1
Speed of movement, m/sec	extremely hazardous (catastrophic)	till 40
	very dangerous	till 30
	Dangerous	till 20
	moderately dangerous	10
Frequency of occurrence, units per year	extremely hazardous (catastrophic)	till 0.01
	very dangerous	0.03–0.1
	Dangerous	0.1–0.2
	moderately dangerous	till 0.1
Avalanches		
Surface area,%	extremely hazardous (catastrophic)	more than 50
	very dangerous	30–50
	dangerous	10–30
	moderately dangerous	less than 10
The area of the manifestation, km^2	extremely hazardous (catastrophic)	more than 5000
	very dangerous	2500 – 5000
	dangerous	1000 – 2500
	moderately dangerous	less than 100
The volume of simultaneous removal, million m^3	extremely hazardous (catastrophic)	3–4
	very dangerous	till 1

Indicators used in the assessment of the degree of danger of the natural process (DNP)	Hazard category processes	Indicators
	dangerous	till 0.5
	moderately dangerous	less than 0.1
Duration, sec.	extremely hazardous (catastrophic)	10–100
	very dangerous	20–50
	dangerous	30–40
	moderately dangerous	till 20
Frequency of occurrence, units per year	extremely hazardous (catastrophic)	less than 0.02
	very dangerous	0.03–0.05
	Dangerous	0.2 – 0.5
	moderately dangerous	till 1

Abrasions and thermal abrasions
The average rate of retreat of the coastline, m /year:

The limits of change	very dangerous	more than 3
	Dangerous	0.4 – 3.8
	moderately dangerous	0.05 – 1.8
Average values	very dangerous	more than 2
	Dangerous	2 – 0.5
	moderately dangerous	Less than 0.5

Processing the shores of water reservoirs
The speed linear coastal retreat in separate sections of the stages of the development process, m /year:

The first	very dangerous	more than 3
	Dangerous	3 – 1
	moderately dangerous	less than 1
The second	very dangerous	1.5
	dangerous	1.5 – 0.9
	moderately dangerous	less than 0.9

Karst

Areal affection of territories,%	very dangerous	5–80
	dangerous	5 – 100
	moderately dangerous	till 5
The frequency of failures of the earth's surface, the number of cases per year	very dangerous	0.1 and more
	dangerous	till 0.1
	moderately dangerous	till 0.01
The average diameter of the holes, m	very dangerous	20 and more
	dangerous	till 20
	moderately dangerous	till 20

Indicators used in the assessment of the degree of danger of the natural process (DNP)	Hazard category processes	Indicators
The total settling of the territory	very dangerous	From low to a few mm/year
	dangerous	Low
	moderately dangerous	low
Suffusion		
Surface area,%	very dangerous	more than 10
	Dangerous	2 – 90
	moderately dangerous	Less than 20
Area manifestations in one area, km^2	very dangerous	till10
	Dangerous	till 5
	moderately dangerous	till 1
The duration of the manifestations of the process, days	very dangerous	till 3
	Dangerous	0.1 -30
	moderately dangerous	more than 10
The speed of the development of the process, days.	very dangerous	more than 10
	Dangerous	more than 0.1
	moderately dangerous	more than 0.01
Subsidence of loess rocks		
Areal affection of territories,%	very dangerous	60 – 70
	Dangerous	50 – 60
	moderately dangerous	30 – 40
Area manifestations in one area, thousand m^3	very dangerous	till 2.5
	Dangerous	till 2.5
	moderately dangerous	till 0.25
Volume subject to deformation of rocks, thousand m^3	very dangerous	till 100
	Dangerous	till 50
	moderately dangerous	till 25
The duration of the manifestations of the process, days	very dangerous	2 – 40
	Dangerous	25 – 400
	moderately dangerous	more than 100
The speed of the development of the process, days	very dangerous	0.5 – 400
	Dangerous	0.1 – 0.5
	moderately dangerous	less than 0.1
The drawn territory		
Areal affection of territories, %	very dangerous	75 – 100
	Dangerous	50 – 75
	moderately dangerous	till 50
Duration of the formation of the aquifer, years	very dangerous	less than 3
	Dangerous	till 5
	moderately dangerous	more than 5

Indicators used in the assessment of the degree of danger of the natural process (DNP)	Hazard category processes	Indicators
The rate of rise of the ground water level, m/year	very dangerous	more than 1
	Dangerous	0.5 – 1
	moderately dangerous	0.5
The erosion of the sheet and gully		
Areal affection of territories,%	very dangerous	more than 50
	Dangerous	30 – 50
	moderately dangerous	10 – 30
The area of a solitary ravine, km^2	very dangerous	0.1 – 3
	Dangerous	0.05 – 0.1
	moderately dangerous	less than 0.05
The speed of development of erosion:		
Flat, m^3/ha /x year	very dangerous	10 – 15
	Dangerous	5 – 10
	moderately dangerous	2 – 5
Gully, m/year	very dangerous	1 – 15
	Dangerous	1 – 10
	moderately dangerous	1 – 5
The erosion of the river		
Areal affection of territories,%	very dangerous	5 – 6
	Dangerous	8 – 10
	moderately dangerous	8 – 10
The length of the coast within the limits of which is relatively at the same time the development of the process, km	very dangerous	200 – 300
	Dangerous	300 – 400
	moderately dangerous	300 – 400
The volume of relatively simultaneous deformations of rocks, million m^3/year	very dangerous	0.2 – 0.3
	Dangerous	till 0.04
	moderately dangerous	till 0.08
The development of the speed, m/year	very dangerous	more than 3
	Dangerous	till 1 – 3
	moderately dangerous	0.1 – 1
Thermal erosion gully		
The potential area affection of territories,%	very dangerous	more than 50
	Dangerous	25 – 50
	moderately dangerous	less than 25
The volume of relatively simultaneous deformations of rocks, thousand m^3/year	very dangerous	1 –10
	Dangerous	less than 1

Indicators used in the assessment of the degree of danger of the natural process (DNP)	Hazard category processes	Indicators
	moderately dangerous	less than 1
The development of the speed, $m^3/m^2 \times h$	very dangerous	more than 0.1
	Dangerous	$0.01 - 0.1$
	moderately dangerous	less than 0.01
Thermal karst		
The potential area affection of territories,%	very dangerous	more than 25
	Dangerous	$25 - 75$
	moderately dangerous	less than 25
Area manifestations in one area, thousand km^2	very dangerous	$0.001 - 1$
	Dangerous	$0.001 - 1$
	moderately dangerous	$0.01 - 1$
The volume of relatively simultaneous deformations, thousand m^3	very dangerous	$1 - 2000$
	Dangerous	$0.1 - 200$
	moderately dangerous	$0.05 - 50$
The duration of the manifestations of the years	very dangerous	$10 - 20$
	Dangerous	5
	moderately dangerous	$1 - 5$
The development of the speed, cm/year	very dangerous	$15 - 100$
	Dangerous	$5 - 15$
The swelling		
The potential area affection of territories,%	very dangerous	more than 75
	Dangerous	$10 - 75$
	moderately dangerous	less than 10
Area manifestations in one area, km^2	very dangerous	$0.01 - 10$
	Dangerous	$0.01 - 10$
	moderately dangerous	$0.01 - 10$
The volume of relatively simultaneous deformations of rocks, million m^3	very dangerous	$1 - 30$
	Dangerous	$0.05 - 1$
	moderately dangerous	less than 0.05
The development of the speed, cm/year	very dangerous	till 50
	Dangerous	$5 - 10$
	moderately dangerous	less than 5
Solifluction		
Areal affection of territories,%	very dangerous	more than 10
	Dangerous	$10 - 5$
	moderately dangerous	less than 5
Area manifestations in one area, km^2	very dangerous	0.0001
	Dangerous	$0.0001 - 1$
	moderately dangerous	$0.0001 - 1$

Indicators used in the assessment of the degree of danger of the natural process (DNP)	Hazard category processes	Indicators
The volume of the unit relative simultaneous deformations of rocks, thousand m³	very dangerous	more than 100
	Dangerous	1 – 100
	moderately dangerous	0.1 – 20
The speed of the development	very dangerous	more than 100m/hr
	Dangerous	from 2 – 10 cm/year
	moderately dangerous	less than 2cm/year
Ice build-up		
Areal affection of territories,%	very dangerous	0.3 – 3
	Dangerous	0.1 – 0.2
	moderately dangerous	less than 0.1
Area manifestations in one area, km²	very dangerous	from 1 – 2 till 50 – 80
	Dangerous	0.01 – 1
	moderately dangerous	less than 0.01
The volume of the unit relative simultaneous deformations of rocks, million m³	very dangerous	1 – 100
	Dangerous	0.01 – 0.2
	moderately dangerous	less than 0.01
The speed of the development, thousand m³/day	very dangerous	5 – 100
	Dangerous	0.1 – 5
Floods		
Areal affection of territories,%	extremely hazardous (catastrophic)	10
	very dangerous	30
	Dangerous	30
	moderately dangerous	70 – 100
The duration of the manifestations of the process, hr.	extremely hazardous (catastrophic)	20 – 25
	very dangerous	1 – 3
	Dangerous	3 – 5
	moderately dangerous	5 – 10
Speed of movement, m/sec.	extremely hazardous (catastrophic)	700 – 100
	very dangerous	50 – 70
	Dangerous	35 – 40
	moderately dangerous	25 – 40
Frequency of occurrence, units per year	extremely hazardous (catastrophic)	0.001 – 0.01
	very dangerous	0.01 – 0.02
	Dangerous	0.02 – 0.05
	moderately dangerous	0.05 – 0.1

85

Sec.5.6. Study of Paleotopography

The third development is connected with the study of the paleorelief and evaluation of its action on the structural integrity of buildings and structures. The studies used georadar complex «Loza».

The aim of the research is to search and localize the ancient paleorelief. In the first place paleowatercourse filled with sediments, as potentially dangerous places when choosing the sites for construction (Fig.5.1). Sedimentary material, filling paleowatercourse, in most cases, is the ideal material for the development of karst-suffusion processes.

In addition, almost all the rivers, including paleowatercourse, or are confined to the faults of the earth's crust, or are folded zones of deformation, with a decrease of relief.

A secondary effect of the paleorelief studies is the detection of increasing the defects of number of trunk pipelines in the places via crossing the paleochannels.

The nature of this dependency on the example of the main gas pipeline (MGP) on the territory of Sverdlovsk region is shown at Fig.5.2.

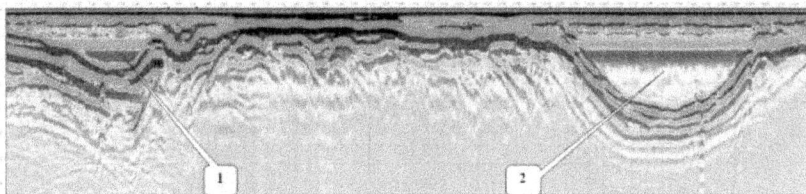

Fig.5.1. Radargram (without treatment), obtained at the intersection of the floodplain of the river Dnepr (Smolensk town): 1 and 2 are the ancient paleochannels of the river Dnepr, filled with river sediments

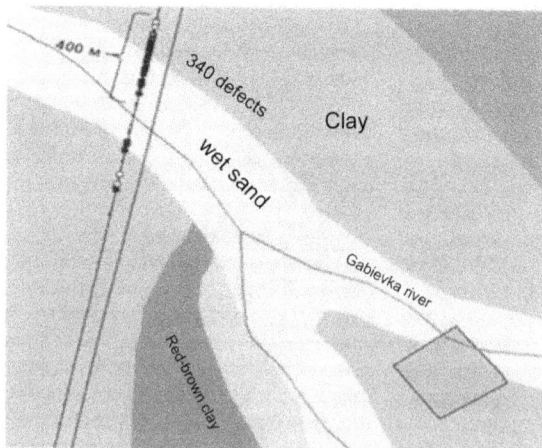

Fig.5.2. The concentration of the MGP defects at the paleochannel intersection

Sec. 5. 7. Anaerobic processes

The fourth development is connected with the study of the problems associated with the effects of sudden collapses of buildings. Some are caused by the occurrence of the concentration in the cavities and detonation of combustible gases, accompanied by an explosion, similar in character of an effect on the design of the volumetric explosion. Studies of this problem are conducted in two directions: development of effective methods of search and localization of soils capable to generate biogas, and study of the nature of the generation of biogas.

In the second direction we study soil gleying - the soil formation process, flowing in the anaerobic reducing conditions with the participation of microorganisms, with the presence of organic matter and permanent or continuous irrigation of individual horizons or the entire soil profile. Question of the study of the processes of the soils gleying in recent years has become more relevant in connection with the specifics of passing these processes in the urbanized territories. It is connected with a complex of problems caused by the bacterial nature of the proceeding processes:

The ideal environment for anaerobic processes are man-made clay loams, underlying asphalt-concrete coating, adjacent to, for example, sewers, with leakage, i.e. in the city, the area of possible distribution of these processes is huge. It should take into account that the speed of propagation of anaerobic bacteria in favorable conditions is very high. So, reproduction, started with one bacterium in ideal conditions, within one year, can play biomass, measuring with a biomass of terrestrial life forms. This is a theoretical analysis. In real terms this can't happen. But filling the pores of the soil with the overlap of flow of ground water is the described fact.

In the process of vital activity of the individual types of anaerobic bacteria involved in the processes of the soils gleying, may emit CH_4, which accumulate in the emerging in the process of gleying cavities, may be the cause of the explosion and sudden collapse of the structures. In the process of loam gleying is the destruction of clay particles, which are binding component of loam, with their transformation in loess and the other is not associated types of soils. This process is associated with a significant decrease of the bearing capacity of grounds and increasing the danger of emergency collapse of the structures.

Ch.6. How to Make Safe the Major Repair

Babayan I.S., Guriev A.A., Iriskulov A.R.,
Kalinin NN, Nikonov T.A., Usatova T.A.[*]

Sec.6.1. General Comments

There is nothing more specific than the object of a major repair. The features of this type of activity require an individual approach. The state of enclosing and load-carrying structures, the characteristics of their engineering solutions, the effect of the environment, at last, a completely different quality of the final product in different houses, all together creates a unique technological methods, identifies specific requirements for the choice of materials; knowledge of regulations, for which it was designed and built by construction, the ability to draw it in a new regulatory «faith». Where is the architecture? Answer: in every detail! In each of the technological subtleties of the amounts of which there is a new quality!

The main properties of the quality consists of the compliance of the functional purpose and the safe operation. Quality is the result of a thorough examination, the competent design and the exact route of the technology.

State institution - the "City coordination of expert-scientific center"-"Enlakom" was established in 1996 with the aim of ensuring the correct selection of materials and technology of the device of the facades. The aim is achieved by the decision of tasks, among which: the development of new materials and technologies for building facades; testing and certification of finishing materials; a survey of facades; the assessment of the documentation; monitoring the construction and operated facades, study and development of domestic and foreign experience. Each of them contributes to the security of the renovated building.

The facade is an element of structure, which at first sight judges the skill of the architect to solve the conflicting objectives of architecture. Shaping, planning, engineering, structural mechanics, acoustics, the ability to fit the new object in the existing building, emphasize the shape, color, material features of its structure, its relationship with the environment all these problems reflected on the facade.

The facade is the relationship between the design, material and technology. Facade is the «face» of the building. It recognized the «disease» of the other parts of the building. If the roof made correctly and the runoff

---

[*] State institution "GUCentr" "ENLAKOM"», Moscow

of rainwater was organized with the mind, it means that the integrity of the facade is guaranteed for many years. Poorly studied the properties of soils, grounds and unexpected rainfall will provoke the emergence of cracks in the walls, etc. The integrity and reliability are the basis of beauty.

In high-rise buildings design of the facades is complicated by many times. Increase of loads, increased wind speeds, thermal toughening of requirements to the structures of the fence, is only part of the list of concerns of designers.

Not all designers know about them. They almost always shied away from this part of the project, giving it the development to the manufacturer, focusing on the cost of the structures and installation. Author's supervision was formal. An abundance of information about the facade systems in various publications rather was the advertising character, than have codified force. The absence of the latter does not allow choose necessary, capable of safe and durable in operation. And along with that correspond to the architectural plans.

Over the last decade, approach to the device facades has changed. As has been said, it is a complex set of activities, strongly influencing the integral perception of the urban space. Diversity is now used facade systems bring to life the special technological and structural requirements, unfortunately still not regulated legislatively. Today in one project can be found the combination of different materials and designs. In this connection there is a necessity to track adopted the draft decision, on the basis of the main tasks: to ensure the assembling organizations of defect-free documentation set. Thus, there is a barrier to the possible errors of the designers in the transition from general ideas to the details of the designing. Today check of project design is not fixed in the standards. Cooperation with design companies allowed, firstly, provide continuously to the designers professional assistance, and, secondly, to form a pool of typical errors.

Sec.6.2. Typical Errors that Occur in the Working Design

In table 6.1 and 6.2 the errors of working design are considered in detail by category.

Listed errors, if miss them into process of production, as a whole, may lead to:
– increase the periods of mounting;
– rising the cost of assembly and operation.

The purpose of the technical surveys and monitoring is to minimize before putting into operation the number of technical and technological mistakes.

Table 6.1

First category. Errors of furnishing and filling out design

№	Typical errors	Possible consequences, if they were not timely identified
1	Working documentation is not being developed and filling out in accordance with GOST R 21.1101-2009 "SPDS. The main requirements for design and working documentation"	Deviation from the approved standards is unacceptable.
2	The project has not agreed with the general designer	The use of solutions is not agreed by the general designer
3	The lack of specifications, general information, registers of working drawings	Using materials that are prohibited to application in the selected system. Add in the working documentation sheets, not passed the coordination
4	The lack of recording of chief project engineer confirming that the design is implemented in accordance with the codes, rules and standards	The chief project engineer did not take upon himself the responsibility for the decisions taken in the working documentation
5	There are no instructions in the working documentation and project of works on the anticorrosive protection of metal elements of exterior facade, used in structures, including their damage in case of the production works	The possible reduction of the term of repair-free service life for facade system.
6	There are not specified tolerances of the distance from the metal goods to the edge of connected bearing elements on the nodes, in the explanatory note to the design documentation and technological cards	Possible collapse of the design of the facade system. Loss of bearing capacity due to the installation of metal goods in the boundary zone
7	The working documentation does not specify the step of fixing work- peace, included in the composition of systems, linking the elements to the bearing structures of the building	Violation of the geometry of the system in the assembly
8	Not specify the attachment of deformation and thermal joints in the facade structures	Possible imperfection, warping of the facing material

№	Typical errors	Possible consequences, if they were not timely identified
9	There are no detailed drawings of the products used in the bearing façade systems (brackets, rails, sliding elements, elements of lining, etc.). In the drawings do not specify the size, material and components (elements) of the framework	There is mismatch of components delivered to the construction site and components from the album of standard solutions. There is the impossibility to manufacture of shaped elements, cornices, etc. There is the discrepancy of the facade elements, designed by the general designer

Table 6.2

Second category. The discrepancy of the design solutions to current standards

№	Typical errors	Possible consequences, if they were nottimely identified
1	Violation of the requirements of fire safety in the places of joining to translucent structures	Fire spread to the whole building. The collapse of the system
2	The wrong choice of material for the walls-ground (in projects of new buildings)	Collapse of the facade system. The deterioration of the energy efficiency
3	Load, design scheme, the safety factor for the anchors were not correctly identified in the static calculation of the frame and anchor elements	Collapse of the facade system or of facade elements
4	Static, dynamic and thermal analysis of structural non-standard units is not fulfilled	Collapse of the facade system. The deterioration of the energy efficiency of the building
5	The nodes do not specify the methods for fixing heat insulation on a supporting base (number, type and depth of the anchoring of dowels)	Insecure attachment of the insulation. The deterioration of the thermal insulation of the building
6	The coefficient of thermal engineering homogeneity of materials and components in applied design is not taken into account in thermotechnical analysis	Deterioration of the energy efficiency of the building

№	Typical errors	Possible consequences, if they were nottimely identified
7	The mismatch of the working documentation and agreed in the "State inspection" the stage «Project» (the thickness of the insulation, material of the bearing wall, the type of facing material)	Solutions which are not agreed upon the General designer. The deterioration of the thermal insulation of the building
8	The discrepancy of working documentation to the passport of color appearance of the object.	The violation of the architectural design
9	Materials of designed structures (including components) will not have passed the evaluation of the suitability for use in the chosen design	Increasing the fire hazard. The reduction in the term of the service facade system without repair
10	The joints of the pair shaped elements (metal castings, "parapet cover", and fire-fighting cutoff) are not elaborated. There is no technological sequence for the device of such joints	Impossible to mount the facade system. Increased fire danger of the facade system
11	There is no technological sequence of the assembly of facade systems In the project of manufacture	Installation of facade system is impossible. Additional costs of material resources and time on rework
12	There are no joints of the interfaces between different systems and structures	The assembly of the system is impossible. Increased fire danger system
13	Lack of protocols of certification tests of fragments in a planned translucent facade system on conformity to requirements of "State ST"	Energy efficiency of the façade systems is not proved
14	The absence of conclusions due to survey of the building before repair	Issue of the disabled project; the probability of oversight of hidden defects, which may appear during the operation

There are some tasks, implementation of which leads to the goal:

1. There is the necessity for detection of structural defects with single surveys and also in time of monitoring of the device or repair the façade system.

2. The fixation of the departures from the standards, technical and technological mistakes made during the assembly of façade system.

3. The preparation of recommendations for the elimination of errors, defects and deviations from the standards.

4. The creation of the bank's common mistakes during installation and defects of facade systems, that occurred during the operation.

Sec.6.3. Typical Errors and Recommendations on their Elimination

Table 6.3-6.6 illustrates the most frequent errors of design and construction works and incorrect operation of the objects.

Table 6.3

Defects and damage of the traditional redecoration

Description	Photo	Recommendations on the elimination
Violation of the tightness of the pairing elements of decorative stucco and plaster surfaces	Photo	Repair coverage, providing a tight pair of elements for the protection of plaster surfaces and stucco from humidity
Traces of moistness on the lower surfaces of the balcony slabs	Photo	Verify the integrity of the insulation layer of the balcony slabs. The lower surface of the balcony slabs dry, pad and re-paint
Delaminating of the first layer of plaster applied to brickwork	Photo	Repair the defective plaster. Remove from the joints the mortar which lost its strength, replace the damaged bricks, restore the lost. Brick surfaces should be rinsed with water under pressure to remove the weak particles and dust, pad them. Plaster with lime-cement mortar (grade 100). Surfaces moisten with water before plastering
Cracks opening width of more than 1 mm	Photo	Make cracks to be widen. Execute repair of plaster of lime-cement mortar grade75. Skim coat (if necessary) to smooth out the rough spots with finished paste-like facade plaster, sanding and willow. Paint the facade surface by paint with given color matching

Description	Photo	Recommendations on the elimination
The absence of a weathering. Destroyed brickwork	Photo	Remove from the joints the mortar which lost its strength, replace damaged bricks, restore the lost. Install a weathering
The destruction of the first layer of plaster applied to brickwork	Photo	For removing of the weak particles and dust from brick surfaces it is necessary to wash them with plenty of water under pressure and to pad. To plaster of lime-cement mortar grade 100, and before plastering surfaces moisten them with water. Pad and skim coat moisture free surface for smoothing wrinkles, sanding and willow. To pad the prepared surface and paint in the color with given color matching

Table 6.4

Defects of hinged facade systems with air-gap

Description	Photo	Recommendations on the elimination
The absence of rigid fixation of bracket on the basis – the attachment of anchor to bracket is not fixing to the designed position		Tighten the anchor manually to the project
In the mounting rail to the bracket hardware installed in the boundary zone of the guide (less than 2, 5Ø from the centre of rivets to the edge of the guide)	Photo	Use an extension to increase the length of the bracket.
Reliability is not ensured and the scheme of fixing heat insulation plates to the base is broken: installed one anchor, in the junction plates. Reduced heat-shielding properties of bearing façade systems –dowels is recessed into the body of the plates damaged outer layer of insulation	Photo	To perform mounting of heat insulation plates of the outer layer of not less than five dowels with the installation in the body of the plates. On the damaged section of insulation set the box (the size not less than 200×200 mm); secure it not less than two dowels
Not eliminated the gap width of more than 2 mm at the joints of	Photo	Impress gaps at the joints of heat insulation plates by the

Description	Photo	Recommendations on the elimination
heat insulation plates. This guide is attached to the bracket of one rivet instead of two.		strips of material insulation on all depth. In addition fasten the real to the bracket with the second rivet
Mobile insert the mounting brackets are not attached to the shelves of the bracket - no rivets. Significantly reduced security operation assembled structures of bearing façade systems	Photo	Install missing rivets.
Slab cladding are shifted relative to the project (the center of the guide), foot fixing the cleat are located in the boundary area of the plates. Reliable fixation of the facing plates to the carrier sub-structure is not providing	Photo	To ensure the project position slab clads relative to the guide, or shift fixing the cleat to the center of the guide in the permissible limits

Table 6.5

Defects of translucent structures

Description	Photo	Recommendations on the elimination
The anchor of bracket is not installed perpendicular to the base	Photo	Set the anchor perpendicular to the base
The anchor of fixing to the bracket is installed with violation of the technologies recommended by the manufacturer of anchors. It is set in the boundary zone of concrete	Photo	Set the anchor from the edge of the concrete at a distance not less than a permitted manufacturer of anchors
Aluminium bracket and member are in contact with cement mortar. Violated the requirements of SNiP 2.03.11-85 «Protection of building structures from corrosion»	Photo	Isolate the aluminium elements of structure of stained-glass panel from the contact with cement mortar
The lack of value of the compensation gap between adjacent to a height of members of stained-glass panel. There is no possibility of temperature displacement of members	Photo	To provide the necessary amount of the compensation gap

95

Description	Photo	Recommendations on the elimination
Adjacent to the height of the members of stained glass rigidly secured to the bracket in place of the compensation gap. There is no possibility of temperature displacement of members	Photo	Perform moving connection of members. To provide the necessary amount of the compensation gap
Cracked glass unit	Photo	Replace the double-glazed window, ensure that the design of stress distribution

Table 6.6

Defects in insulation systems with a thin outer plaster layers

Description	Photo	Recommendations on the elimination
Gaps at the joints of heat insulation plates filled with glue composition		To remove the glue from joints of plates. Fill in the gaps strips of material insulation on all depth
Violated the scheme of installation of fire-prevention baffler of mineral wool plates at the corners of the window openings (joints of heat insulation plates coincide with the lines of the corners of the aperture), around the emergency exits (the width of the transverse cross-section baffler <1.0 m)	Photo	Perform installation of mineral wool plates at the corners of openings with a corner cut in the plates in place, with the distance from the edge of heat insulation plates to the corners of the aperture not less than 200 mm. To ensure wide bafflers around escape exit ≥ 1.0 m
Infringed the provisions of the fire safety in case of installation of fire-prevention bafflers: the adhesive should be applied continuous layer	Photo	Perform installation of bafflers, to ensure the continuity of the adhesive layer between the insulation and the basis
Broken technology of device of the base reinforced layer: the adhesive is applied through reinforced grid deposited with "dry" way on the insulant	Photo	Remove plaster layer to the insulation. Perform basic unit layer according to the requirements of code-technical documentation. This reinforced mesh must be located in the middle or in the upper third of the base layer

Description	Photo	Recommendations on the elimination
Broken technology unit of the base of reinforced layer: the adhesive is applied through reinforced grid laid on the "dry" insulate	Photo	Remove plaster layer to the insulation. Perform basic unit layer according to the requirements of code-technical documentation. This reinforces mesh must be located in the middle or upper third of the base layer
Breach of the scheme of application and the area of the adhesive composition (less than 40% of the slabs). Not executed priming the surface of the base	Photo	Complete dismantling of the insulation. Preparation of the foundation. Re-installation of insulation with observance of requirements of code-technical documentation

Sec.6.4. Energy-Saving as an Important Component of the Major Repair

Saving energy – is the most important national problem of today. The technological development of the country, as a rule, highlights the energy-saving programs, which, in turn, closes with a system of measures, which supports the ecological balance of the natural environment and human activities of the society. Two "e" – economics and ecology - lie in the basis of the state approach to the management in the countries of the European Union. That is why the share of energy costs in the cost of, for example, reinforced concrete products in the Netherlands do not exceed 1% against 10–15% in Russia. Therefore the economy of material resources, and first of all, energy is the main task. Saving energy can be in everything and everywhere, inclusive of major repair too.

Getting to a comprehensive overhaul one will need to imagine what needs to be done. There are a lot of problems that need to be solved, most of them would increase energy efficiency. For example – there should be winteraztion of external walls insulation, windows and balcony doors should be replaced. As soon as the apartment has a new look and the walls shine with new plaster or tiles, immediate next step would be glazing of balconies and loggias. Then it should be used to insulate overlap cold attics and replace the roof (where it is necessary), update the engineering networks, service points and the equipment inside of the houses, install water and heat meters. The data published by AVOK ("Automation, water supply, heating, air conditioning" Publication) and the Department of Major Repairs of the Housing of Moscow Fund, demon-

97

strates thatthe cost savings from improved thermal insulation of walls and ceilings in cold attics is equal to 25%; installation of new windows - 10%; installation of automated control nodes device in systems of heating and location of thermostat on the heating appliancesis another 18%. And, finally, the reduction of excess of interchange of air in the apartments is 6%. It is known that all the residential houses, schools, kindergartens, hospitals, clinics and other objects of social sphere, built in the 60-ies until the end of the XX century, now don't comply to the current standards for thermal building protection. In terms of quantitive indicators there are more then 100 million square meters of residential housing. And there are also about15 million square meters of the general educational institutions, and about 9 million square meters of outpatient clinical institutions. And all of those buildings require implementation of thermal protection.

Therefore, economic component in the repair of the facades is no less important than the aesthetic one when 500k to 1800k rubles annually can be saved for each building. Multiplying Russian rubles on the number of houses we will get a huge cost savings, fuel savings, and also, no doubt, improvement of air environment in town. So the special value gain search of new thermal insulation materials, their testing in the natural conditions. Physical-mechanical properties of porous heat-insulating sanitized plaster "TEPLOSAN", developed in "Enlakom" and experimentally verified during the overhaul of apartment houses in Moscow and other similar plastering compositions («Thermo-Eco» produced in Turkey, "BOLARS, Easy Wall" made in Russia), show that the warming of facades in this way, though won't solve all the problems of energy-saving in housing and communal sphere, but, undoubtedly, should be the first and mandatory step in the right direction.

Sec.6.5. Façade Systems Requirements

Facade systems (FS) require:
– evaluation of project design, selected materials;
- presence of the technical certificate and regulations.
The monitoring service shall verify:
– evidence of the durability of the FS, comparable with the building working life;
– corrosion resistance of structures;
– fatigue strength of metal structures;
– the results of the aerodynamic research;
– maintainability of the system and preventive care for it;

– the impact of initial imperfections of the FS grounds at its subsequent operation;

– obligation of the tuning works, leading the base of FS to the standard.

Features of major repair create difficulties in achieving regulatory thermal insulation.

It would be to ration the energy efficiency of the residential and other buildings after the major repair in the specific consumption of heat energy for heating. It should be installed in each design the maximum value of this indicator in the terms of reference.

In should be also taken into account that the feedback of tenants regarding microclimate in the apartments after repairs as an important quality indicator of a major overhaul.

Ch.7. Buildings Structures in Complex Conditions of Operation

*Banach V.A., Banach A.V.**

Sec.7.1. General Comments

In the operation of buildings and structures in complicated engineering-geological conditions almost always take place deformations of the ground under the foundation , the consequence of which is the deformation of the buildings. Studies show that, in such circumstances, the majority of buildings belong to the deformation state. In case of dynamic effects such strain state of the buildings, which can be considered as preliminary, affects their behavior and changes the stress-strain state of the individual structural elements and buildings altogether.

It is known that the dynamic actions are defined as free dynamic characteristics of the buildings, so and are forced, and also the parameters of their stress-strain state. However, the behavior in the process of exploitation of the pre-deformed building will be different from its behavior, expected in the design. You need to give a quantitative and qualitative assessment of the influence of the deformed state on the dynamic characteristics of the buildings under seismic impacts.

Part of the objectives of this research are the identification and analysis of features of formation of computational models due to operated buildings, which are deformed as a result of the effect of irregular settlements of the ground, when they are exposed to the dynamic effects. The peculiarity of the long-term operation of buildings and structures is result of deformation of the ground. They get non-uniform settlements, which become the cause of change of altitude position of the load-bearing structures, their banks, non-uniform settlements of the reference sites, presence of defects in the form of cracks, chips, outcrops of the working reinforcement in reinforced concrete structures and other things. To the greatest extent such phenomena are typical for buildings which are operated in difficult engineering-geological conditions, but they can also take place under normal conditions of operation.

Thus, the technical state of building being in the complex of engineering-geological conditions is characterized by close to be exhausted or expired service life, the presence of stress and strain caused by non-uniform settlements of ground, the presence of load-bearing structures in limit state, spontaneous reconstruction and redevelopment of operated buildings.

---

* State higher educational establishment «Zaporizhskya state engineering Academy», Zaporozhye, Ukraine

Sec.7.2. Deformation Model

For all groups of models their deformed scheme is used taking into account possible (in the design and construction) or factual (in the operation and reconstruction) deformations, caused by non-uniform settlements of grounds. There are three possibilities to obtain the deformed model:

– an adjustment of the geometric scheme of the designed model according to the data of full-scale survey (for exploited buildings) or on the results of the calculation of the biases and banks (for the future buildings);

– a preservation of the deformed shame of the designed model in a finite elements method, obtained as a result of static calculation of the system "building – foundation", as the initial data for the next stage of the calculation on a specially designed algorithm;

– a transformation of the displacements of the nodes of the model obtained in the calculation of the system "building – foundation", into the equivalent load, implemented in the software of finite element complexes (for example, LIRA-Windows).

This way of specifying the deformed schema of the building (structure) allows to take into account the background of its loading, when the dynamic response of structures of the building is superimposed on the stress state, arising as a result of non-uniform deformations of soil grounds.

The design of such buildings and structures, especially for dynamic effects, is a complex scientific and engineering challenge. The difficulty lies in the need to establish an adequate mathematical model, with the parameters of static and dynamic work, which will be qualitatively and quantitatively coincide with the same parameters of their natural livework (of course, in the acceptable range of adequacy).

The carried out researches have shown, that the interaction of buildings with underground bases significantly affects to the static and dynamic characteristics of the work of the buildings, so, at first glance, computational models, which take into account the interaction of buildings on the bases, are more appropriate. But poorly or carelessly prepared detailed calculation model can give a result far from correct. It should also take into account the fact that in the regulation documents (for example, [7.1, 7.2]) there is no clear guidance on how to use the simulation models for dynamic analysis, including the ways of accounting for the interaction with the ground bases.

Considering the specifics of calculations of the buildings at the dynamic effects and the limited number of experienced specialists in these field, necessary engineering techniques is needed, which will allow even to the engineer with the entry level design experience to form a calculation model that is acceptable for dynamic calculations, and also to confirm its adequacy.

The most fundamental, however, remains the question of the necessity of considering the actual strain state in the calculations of operated buildings on dynamic effects, since the accounting of the compliance ground significantly affects the dynamic characteristics and stress-strain state of building structures.

On the one hand, the such design model, in which the ground bases presented by the spatial finite elements considering the physical and geometric nonlinearity, provide an opportunity to carry out calculations of the interaction of buildings with the grounds for any engineering-geological conditions. The problem of accounting for pre-strain state of the building is solved automatically with the help of correct modeling of soil array (distribution of rigidity characteristics according to the engineering-geological surveys, irregularities in the ground bases, underground structures and communications, the effect of the surrounding buildings and other factors).

On the other hand, any unjustified complexity of the model leads to an increase in the probability of errors of principle, and simplifications, suitable for static calculations of buildings and structures, are categorically unacceptable for dynamic design.

Finally, at this moment software applications have not solved the problem of separate assignment of the values of the vibrations logarithmic decrement, which negates the attempts to adequate assignment of combined systems from different materials and subsystems (including the basis) for dynamic calculations. The exception can be considered a seismic pulse (explosive) impact, since their effect on the building close to instantaneous, when the system "building - the ground base" responds to such dynamic load with no significant difference in the basic dynamic characteristics (logarithmic decrement vibration, dynamic strength, rigidity and the like). The exception is high-rise buildings and buildings with large spans.

Sec.7.3. Structural Diagram and Dynamic Actions

For correct account of dynamic effects on the building structures it is necessary to form their detailed spatial structural diagrams with the basic structural elements, and, in contrast to the static model, dynamic model should include the load bearing and self-bearing elements (under condition of setting the correct way their interface with supporting structures). It is desirable to take into account the flexibility of the joints, physical, geometrical, and structural and genetic nonlinearity.

Also an important factor in the formation of computational models of buildings and structures, which are operated in complex engineering-geological conditions due to the dynamic actions, is an account of their deformed structure, which includes stress and strain accumulated in the process of operation with non-uniform deformations of the base. The numerical experiment was held for the building, represented by two models of interaction with the foundation. There were simulated non-uniform deformations of the base, and, based on the results of these calculations, replacement of movements with equivalent loads was made by the using special function LIRA-Windows software, which adjusted the .design of a building to create a model of its strain state.

For each of the two variants of interaction of the building with a ground, taking into account the soil-base special finite elements and in the form of soil array represented by three-dimensional finite elements - both types of models, the original and a strain were calculated on the seismic actions. While comparing the results of the calculations provided the identity of dynamic effects (analyzed parameters of oscillation of the structures, as well as of their stress-strain state). Computational models for the initial and strain variants of calculation are shown in Fig.7.1.

Detailed spatial design finite-element model of operated buildings – large-panel ordinary block-section model (the series I-480A), with regard to the non-uniform deformations of the foundation from subsidence, which were set according to the results of full-scale survey, were adopted as the object of study. This way of specifying the deformed structure of the building allows to take into account the background (where we have jointly geometric nonlinearity and "genetics") of its loading, when the dynamic response of structures of the building is superimposed on the stress state, which has arisen as a result of uneven deformations of soil-grounds, and which, in turn, leads to the increase of stress in a structural elements in comparison with not deformed scheme.

The aim of the numerical experiment was a comparison of the dynamic characteristics of the model and the parameters of their stress-strain state for quantitative and qualitative evaluation of deformed scheme, as well as possibilities of use of such models in engineering dynamic calculations. Seismic action with the intensity corresponding to 6 points also adopted as a dynamic effect.

Frequency spectra of the lower forms of natural oscillations of the building, the share of which was the maximum used modal of mass of the system, as well as the numeric values of the dynamic characteristics for each of the options, are given in the table 7.1, and their comparison can be found in the table 7.2. The systems with finite-element model of the soil (Fig. 7.1c, d) were used to compare the dynamic characteristics of the values of deviations from the values of the reference.

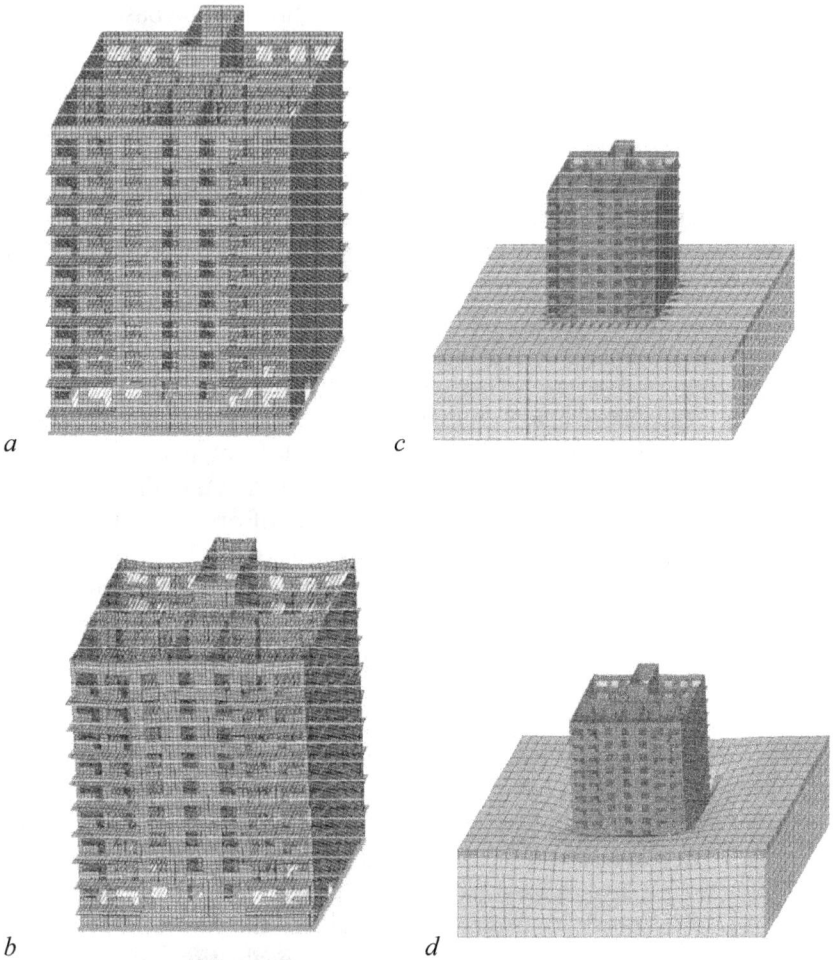

Fig.7.1. Calculation model of interaction of buildings with the bases due to the dynamic actions: a – records of the soil-base special finite elements, the original scheme (option 1); b – the same deformed scheme (option 2); c – in the accounting base in the form of soil array, presented by volume of finite elements, the original scheme (option 3); d – the same, deformed scheme (option 4)

Table 7.1
The dynamic characteristics of computational models of buildings with the light foundation (the original and deformed scheme)

Variant	Frequency spectrum of fluctuations, Hz	Forms of natural oscillations			
		1	2	3	4
1		0.191	1.594	2.809	5.281
2	Гц ├─┼─┼┼─┼─┼─┼─┼┼─┼─► 0 1 2 3 4 5 6 7	0.182	1.476	2.630	5.483
3	Гц ├─┼┼┼─┼─┼─┼─┼─┼─┼─► 0 1 2 3 4 5 6 7	0.668	0.882	1.165	1.451
4	Гц ├─┼┼┼─┼─┼─┼─┼─┼─┼─► 0 1 2 3 4 5 6 7	0.616	0.854	1.055	1.397

Table 7.2
The deviation of the characteristics of natural oscillations of the design deformed model for the building with a view of the grounds

Variant	The dynamic response	Deviation,%, characteristics of forms of natural oscillations			
		1	2	3	4
2	Frequency of oscillation	-4.9	-8.0	-6.8	3.7
4		-8.4	-3.3	-10.4	-3.9

Parameters of the stress-strain state due to loadings were analyzed for different models. Numerical values for each of the options are shown in table 7.3, their comparison is shown in the table 7.4. The move of the nodes of the system was analyzed mainly according to the forms of the oscillations as the main indicator of the differences of the studied models. It should be noted that the frequency characteristics for models with regard to soil-dimensional array elements close to the ones recommended by international standards [7.3].

The results of the comparison of the parameters of the stress-strain state, namely, the movement of the nodes of the estimated model shows, that the preliminary deformation under dynamic calculations of buildings and structures leads to an increase in movements to 12.5-33.3% for models with regard to soil-grounds of special finite elements, and by 6.0-12.5% for models with regard to the foundation in the form of soil array represented by three-dimensional finite elements.

Similar calculations are made for computational models of other types of buildings (the heighten storey's building with a monolithic reinforced concrete skeleton and a 5-storey frameless residential brick building with long term of operation).These calculations clearly confirmed the tendency of increasing movement while keeping the initial deformations. This is especially important for operating buildings affected by non-uniform settlements, when accumulated in the process of exploitation of deformation close to the limit.

It is necessary to take into account the deformed scheme in the estimated models in case of taking into account the above arguments, with the possibility of appearance of uneven ground deformations of designed buildings and occurrence of preliminary deformation of operated buildings.

Table 7.3

Characteristic displacements of nodes by the calculated model of the building with the account of the foundation (the original and deformed scheme)

Variant	Direction	The maximum displacement due to loadings, mm							
		1	2	3	4	5	6-1	6-2	6-3
1	On the axis X	21.1	0.2	0.1	20.7	20.6	2.3	1.1	0.7
	On the axis Y	68.7	0.3	0.1	36.9	36.8	4.1	0.8	0.2
	On the axis Z	43.5	1.1	0.1	1.3	1.3	0.2	0.2	0.2
2	On the axis X	23.4	0.2	0.1	22.9	22.9	3.1	1.3	0.8
	On the axis Y	69.5	0.3	0.1	38.4	38.4	5.0	1.0	0.2
	on the axis Z	43.6	1.2	0.1	1.5	1.5	0.3	0.3	0.2
3	On the axis X	7.4	2.1	0.1	0.1	0.1	0.1	5.4	2.9
	On the axis Y	7.1	6.2	0.5	4.5	4.5	2.7	0.3	0.2
	On the axis Z	148.0	14.0	0.8	1.0	1.0	0.6	1.7	1.6
4	On the axis X	7.5	2.2	0.1	0.1	0.1	0.1	5.4	3.0
	On the axis Y	7.5	6.3	0.5	4.7	4.5	2.8	0.3	0.2
	On the axis Z	149.1	24.7	0.8	1.0	1.0	0.6	1.7	1.6

Table7.4
The deviations of the displacements values of characteristic nodes of the design model due to deformed building with a view of the grounds

Va-riant	Direc-tion	Absolute deviation,%, displacements values due to loadings							
		1	2	3	4	5	6-1	6-2	6-3
1	On the axis X	9.8	0.0	0.0	9.6	10.0	25.8	15.4	12.5
	On the axis Y	1.2	0.0	0.0	3.9	4.2	18.0	20.0	0.0
	On the axis Z	0.2	8.3	0.0	13.3	13.3	33.3	33.3	0.0
2	On the axis X	1.1	1.8	8.3	8.3	9.1	12.5	0.6	1.0
	On the axis Y	1.1	1.3	3.8	4.1	0.5	1.4	3.9	6.3
	On the axis Z	0.7	4.8	2.5	0.0	1.0	3.1	1.2	0.6

At reconstruction of buildings machinery and equipment that are be-ing usedcause the dynamic impact on the structures. These impacts can be attributed to the category of impacts of low intensity, and although they, at first glance, do not have a significant influence on the strength and stiffness of the bearing constructions, and their influence on people presenting in the premises may be extremely negative.

It is known that the person is negatively affected by certain wave band, causing a negative reaction of the organism and regulated by sani-tary regulations. Method of direct instrumental measurements of dynamic characteristics during operation of the equipment gives the opportunity to control the compliance with the standards, but in this case there is no pos-sibility to predict the negative influence of dynamic effects on people, regardless of the spatial position in the building. The possibility of simu-lating such processes allows to access risks forecast development of negative processes in time.

In addition, the negligible contribution of dynamic effects of low in-tensity in the limit load for the structures is typical for the buildings, op-erated under regular and favorable conditions. In those cases, when the individual structure or a building is in the limit state or close to it due to operation in complicated engineering-geological conditions, even a small increment of the forces and stresses in the structural elements can lead to exhaustion of the service life of the building.

Dynamic effects of low intensity, particularly in the case of their permanent influence that lasts for years and decades, will cause local fatigue destruction of materials of structures, which can also cause a reduction in the service life of the building. Thus, the required quantitative and qualitative assessment of the actions of this type on the structural elements of operated buildings.

Indicators of comfort for people in operating buildings directly depend on the frequency spectra of the response of the elements of these buildings under the dynamic impacts, including the sound waves (indicators of the level of noise)and wind exposure. The analysis of the dynamic responses of buildings and structures on these actions is also necessary to provide the required standards of comfort.

Sec.7.4. Analysis of the Dynamic Responses of Structures

The dynamic actions of low intensity for exploited buildings can be classified as following:

– technological actions from construction equipment and tools used in the reconstruction and repairs (for example, punching of openings in walls, plates of floors, the dismantling of elements, the restoration and reinforcement of bearing structures, and so on);

– technological actions from household and industrial equipment that does not meet the intended purpose of the building (industrial air conditioners, refrigeration equipment, ventilation and compressor units and the like in the facilities under their re-profile);

– sound effects from external sources of noise, including those outside distances taken into account due to the transfer of impacts through the ground, causing the dynamic response of structures of the building or the discomfort of a person in this building (aircraft, railway, heavy-load transport, explosions in mines, hands-free work of the equipment in close proximity to the building and so forth);

– wind actions to high-rise buildings, when it is necessary to make a frequency analysis to ensure the requirements of comfort and strength.

In determining the effect of the vibrations from construction machinery and equipment on the human body it is expedient to fully investigate the whole design model of the object, as it is necessary to analyze the whole spectrum of amplitude-frequency characteristics as the building as a whole, as well as its separate structures because including sanitary regulations on frequencies, that can cause a negative reaction of the human body. Studies have shown that the construction and repair equipment, used for the reconstruction and repair of buildings, may cause actions safe for the building structures, but harmful for human health.

For research of dynamic reaction of building structures on the impact of technological equipment consider reconstructed residential five storeys building in the Zaporozhye city, where on the first floor after the enlargement of openings in bearing walls and re-profiling of the premises there was established Dry Cleaning/Laundry. On the floor above the cellar, equipment with cyclical engines (frequency of up to 1000 turnover per minute.) has been set up. In addition, on the fourth floor of the residential apartment there reconstruction took place and the new balcony was built as well as the openings in the exterior walls were expanded. The building was frameless, with load-bearing exterior walls of silicate bricks, and with internal walls carrying out of the clay brick. The rest of the building consisted of plaster-concrete, the floors had reinforced-concrete hollow-core slabs and the foundation – tape-monolithic, reinforced concrete and the substrate preparation created from 2.2 m thick dirt pillows.

Finite-element calculation model of the building includes the load-bearing walls, partitions, floor slabs, foundation. The base is considered with special finite elements modeling unilateral elastic constraints.

Calculations were performed using the software LIRA-Windows version 9.4 (license NIIASS № 1D/549 for ZGIA № 9Y037014). According to the results of calculation design combination of forces and principal stresses in the bearing structures are defined. The module LITERA of this software was used to determine main stresses. The calculations were based on force design combinations. The criteria of theory maximum principal stress have been used.

The deformed scheme of the building was taken into account, obtained as a result of his examination, and partial reconstruction, associated with the device for the new and expansion of the existing balconies, as well as the expansion of the window openings in the level of the fourth floor. In determining the strength characteristics of the brickwork the age was taken into account, as well as the presence of aggressive components of the environment and the presence of defects.

The distribution of the main compressive stresses in the external walls of the static load showed maximum value 960 kN/m^2. This value is less than the limit for brickwork considering its age and the presence of defects 1020 kN/m^2.

In the calculation of the dynamic impacts of technological equipment on the ceiling of the first floor a harmonic load from the additional weight of 1.0 kN with a frequency of 16.7 rev./sec. (1000 revolutions per min) was applied for the place where this equipment is installed. The dynamic load was attached also, simulating the work of punch with the expansion of openings, like the pulse (shock) loads from the additional weight of 0,5 kN and repeatability of 5 impacts per second, in the levels of the piers of the first and fourth floors. The calculation was carried out to determine the amplitude-frequency characteristics and parameters of the stress-strain state structures, as well as in the time domain to determine displacements, velocities and accelerations due to vibrations.

Increment of the main compressive stresses in the places of their maximum values of static loads were obtained in the amount of 93-137 kN/m^2, when determining the parameters of the stress-strain state of structures from the dynamic effects. Thus, the total main compressive stresses are 1053-1097 kN/m^2, which exceeds the strength of masonry equal to 1020 kN/m^2. Hence, the work of technological equipment in the building will lead to a limit states in the brick wall, with the most vulnerable are the places where the load-bearing structures of the device of the new balcony are being fastened, which means inadmissibility of the spontaneous reconstruction of buildings, operated in difficult engineering-geological conditions.

The calculation is made for the same model in an interval of 10 seconds. Design displacements, velocity and acceleration from vibrations in the directions for the control points of the estimated model are shown in Fig.7.2 and 7.3.

Comparison in time of technological equipment activity the dynamic characteristics of structures are given in the table 7.5, the frequency spectrum of fluctuations in Fig.7.4.

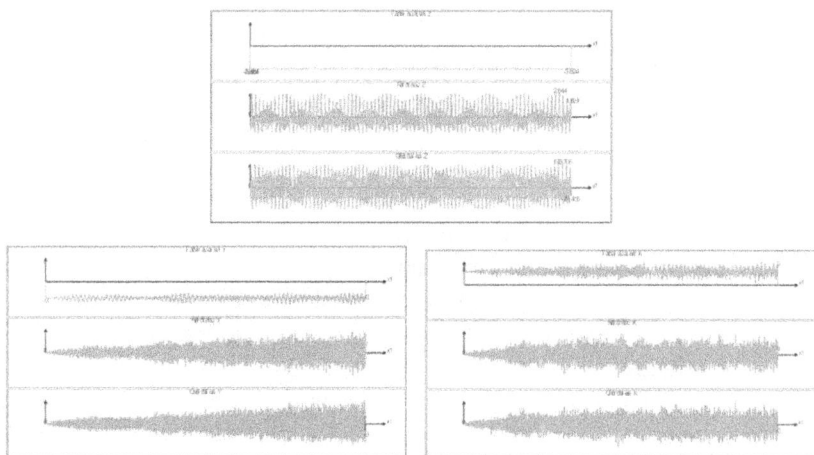

Fig.7.2. Diagrams of vibrations of the floor slabs on directions -displacement, speed and acceleration, in the interval 10 sec. at work of centrifuge (16.7 rev. per sec.)

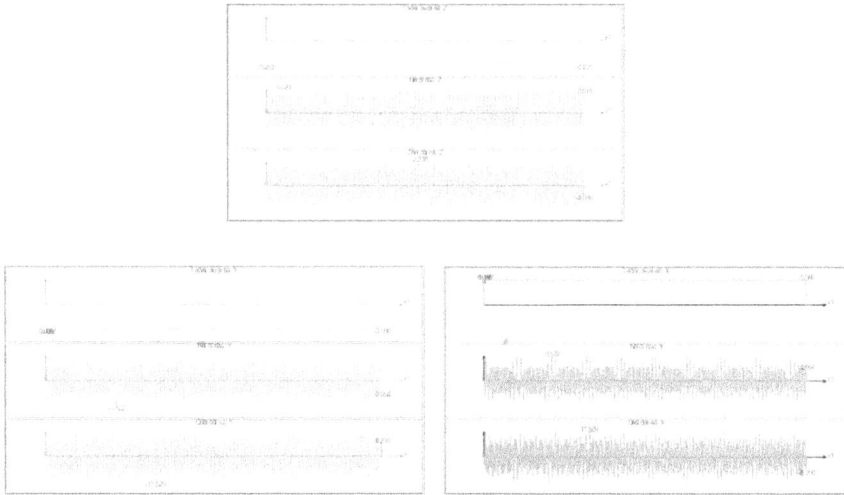

Fig.7.3. Diagrams of vibrations of the outer wall on the directions – displacement, speed and acceleration, in the interval 10sec at work of perforator (5 impacts per sec.)

Analysis of the results of the calculation, submitted for the control points of the ceiling of the first floor when working centrifuges and the pier of the fourth floor with the expansion of openings with perforator, and their comparison with the results of full-scale measurements (see table 7.5), allow to make a conclusion about the proximity of the dynamic characteristics of the model and the real object. The deviations of the parameters are a 3.0-13.8% for the amplitudes, 9.5-17.2% for velocities, and 2.3-30.3% for accelerations.

In addition, there has been an excess of the permissible acceleration for the vertical oscillations of the work of centrifuges, which according to the results of full-scale measurements is 204, 35 mm/sec^{-2}, the results of calculation is 199, 70 mm/sec^{2} while the maximum allowable is 150 mm/sec.^{2} It shows about the violation of sanitary requirements and the need to use special measures for the damping of the oscillations of working centrifuges.

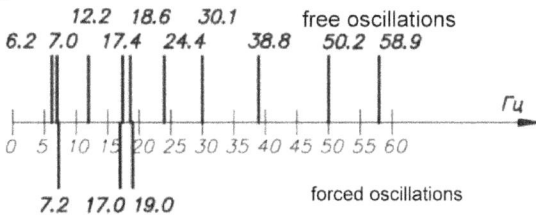

Fig.7.4. The frequency spectrum of free and forced oscillations of computational model of a building with technological impacts

111

Table 7.5

The dynamic characteristics of the control points of the building according to the results of full-scale measurements and calculations

Actions	Oscilla-tions	Fre-quency, Hz	Amplitude		
			Displace-ments, mkm	Veloci-ties, mm/sec.	Accelera-tions, mm/sec$^{\cdot 2}$
Centrifuge (measure-ments)	Vertical	7.00	54.00	2.370	204.35
	Horizon-tal, on X	–	–	–	–
	Horizon-tal, on Y	+	–	–	–
Perforator (measure-ments)	Vertical	0.15	20.00	0.019	1.78
	Horizon-tal, on X	2.50	6.50	0.102	16.02
	Horizon-tal, on Y	2.60	6.20	0.101	16.53
Centrifuge (calculation)	Vertical	7.20	58.24	2.644	199.70
	Horizon-tal, on X	17.00	0.00	0.00	0.00
	Horizon-tal, on Y	19.00	0.00	0.00	0.00
Perforator (calculation)	Vertical	0.16	23.20	0.021	2.24
	Horizon-tal, on X	3.00	6.70	0.122	11.53
	Horizon-tal, on Y	3.00	6.70	0.122	11.53

Thus, the detailed and full accounting of the calculated models for dynamic effects of low intensity allows obtaining the characteristics, influencing not only on the strength of structures, but also on the human body. Account of the peculiarities of the estimated models enables also to predict negative dynamic effects in the reconstruction.

Ch.8. Study of the Regularities of Structural Physical Wear due to the Buildings of Thermal Power Factories

K.I. Eremin, S.A. Matveyuchkin, G.A. Pavlov,
*E.L.Alekseeva, A.N. Shuvalov**

Sec.8.1. General Comments

Enterprises of power engineering are strategically important objects in the Russian Federation. The occurrence of accidents and emergency situations on these objects may be not only local, but also global in nature. It is very difficult and sometimes impossible to fully predict the economic damage and the consequences of accidents in these enterprises.

The sphere of energy is a closed and consequently poorly understood. There is no generalized data, at this moment, on the failure distribution of the enterprises of power main buildings frameworks elements , taking into account the duration of their operation.

Analysis of a small number of the main buildings of the enterprises of power engineering showed that the buildings are in a satisfactory condition, but the condition of many of bearing structures have been assessed as limited efficient.

The main building of the thermal power plant is a building or complex of buildings in which the basic and auxiliary equipment is placed and is directly involved in the generation of thermal and electric energy [8.1]. The main building receives the fuel subjected to incineration, cold water for cooling of waste steam and other purposes. From the main building the warm water is being flown after from capacitors, as well flue gases, slag, ash draw and thermal and electric energy.

The most complex and expensive equipment is concentrated in the main building, the value of which, as a rule, is more than half the cost of the whole complex of objects of thermal power station.

The layout of the main building should meet the requirements for the placement of technological equipment, which contributes to reliable and uninterrupted power supply for consumers, the efficiency of operation of the power factory as a whole, and satisfaction of its high technical and economic indicators.

The layout should provide:

– conditions for inspection, repair, assembling and dismantling of equipment;

* Ltd "WELD" Co. Magnitogorsk, Chelyabinsk region, Moscow State University for Civil Engineering, Moscow

– the necessary sanitary and hygienic conditions of work for maintenance and operating personnel;

– fire and explosion safety;

– safety of personnel, protection of the environment in case of accidents at power plants and in extreme natural influences;

– high technical and economic indicators of the main building;

– opportunity for reconstruction of the completion of the design lifetime of the equipment, as well as the complete dismantling with the restoration at the site of the primary natural conditions.

Sec.8.2. Analysis of Structural Decisions

The analysis of structural decisions was carried out in eight main bodies of of private electricity and heat generating company Ltd "Fortum". The installed capacity of "Fortum" consists of the electricity more 2785 MW, and of thermal energy 11862gral/hr.

"Fortum" objects are located in the Urals and in Western Siberia. In the network of the company there are eight thermal power plants: 5 of them - in the Chelyabinsk, 3 - in the Tyumen region.

The main buildings of "Fortum" were put into operation at different times, from 1931 (Chelyabinsk GES) to 1996 (Chelyabinsk TES-3).

On the basis of the analysis of structural solutions of the main buildings of the enterprises of the heat-power engineering, the following result was established: in the framework of structural decisions of main buildings laid multispan frame, which includes not less than three spans. The design of the main building is determined by the structural scheme of some of its spans. Spans are performed entirely in reinforced concrete, metal frame, or mixed with reinforced concrete columns and metal structures of the coating. The framework consists of columns, beams and trusses, forming in the transverse direction frame with rigid or hinge joints. The rigidity and stability of the frame and its individual elements are provided with the system of ties: vertical ties on columns, perceiving longitudinal forces from the wind actions on the butt-end of the building and the forces of the longitudinal deceleration of cranes, and also by horizontal and vertical ties in the tent roof of the building, providing the stability of the coating.

The research covers buildings with 6.5–45.0 m spans.. Width of the crossings of the main buildings varies and depends on the destination. Minimum span width is 6.5 m (bunker branch); maximum width is 45.0 m (engine and turbine department). Step columns on the main buildings varies and is in limits from 6.0 up to 13.0 m.

The main buildings are equipped with overhead cranes with carrying capacity up to 100 t, working at heights of up to 60, 0 m.

Features of structural decisions of main buildings of the Chelyabinsk GES (year built 1930) and the Chelyabinsk TES-3 (the year of construction of 1996-2006), and the comparison of the main parameters of the main buildings is presented in the table 8.1.

Table 8.1

Comparison of the basic parameters of the main buildings

Title	Stage of constr uction	Year of putting into opera- tion	Depart- ment	Span width, m	Span height (to the bottom of roof truss),. m	Step of columns	Step of roof trusses (roof beams), m	Structural scheme
Chelya- binsk TES-3	I turn	1996	Machi- nery	45.0	27.55	12.0	12.0	Metal skeleton
			Deaerator- tank	12.0	41.3, 38.3	12.0	12.0	Metal skeleton
			Boiler	42.0	64.55	12.0	12.0	Metal skeleton
	II turn	2006	Draught unit de- partment, (frame with 5 spans)	60.0	12.30	12.0	12.0	Metal skeleton
Chelya- binsk GES	–	1930	Machi- nery	17.25	20.6	6.5,6.7	6.5,6.7	Rein- forced concrete skeleton
			Boiler	23.0	29.8	6.5,6.7,13. 0	4.35,4.5	Mixed skeleton
			Bunker	6.5	32.0	13.0,13.2	13.0,13. 2	Rein- forced concrete skeleton

Analysis of the key load-bearing structures of the main buildings is represented in table 8.2.

On the basis of the structural schemes of the two main buildings hydroelectric and thermoelectric power plants, built in the beginning and at the end of the last century, there can be seena transition between using reinforced concrete structures in the frame of the building and using metal. Currently, the roof is mainly performed with lightweight of profiled steel sheets or panels of «Sandwich» type, which greatly reduces the load

115

transmitted to the supporting structures of the building. Overall dimensions of the main building of the Chelyabinsk TES-3, erected in the later period, significantly exceed the width and the height of the main crossings of the dimensions of the main building of Chelyabinsk GES, erected in the beginning of the last century.

Table 8.2

Analysis of the based bearing structures of the main buildings

Title of department part-ment	Cross-section of column, mm		Material of column	Bearing structures of covering	Span, m	Material
Main building of Chelyabinsk GES						
Boiler	Rectangular 1000x1400 Height 29.8m	Rectangular 550x1150 Height 29.8m	Reinforced concrete Class B25	Metal roof trusses trapezoidal shape. Truss elements of the compound t-cross-sec., 2-formed equal parts	23.00	Steel grade St3ksp
Machinery	Rectangular 700x120 Height 19.8m	Rectangular 700x1250 Height 19.8m	Reinforced concrete Class B25	Monolithic reinforced concrete beam coating I- cross-section 900x700mm	17.250	Reinforced concrete class B25
Main building of Chelyabinsk TES-3						
Boiler	Height ,60m	Height,60m	Steel Vst3sp5	Metal roof trusses trapezoidal shape. Truss elements of the compound t-cross-sec., 2-formed equal parts	42.0	10D2S12
Machinery	Height 27.55m	Height 27.55m	Steel Vst3sp5	Metal roof trusses trapezoidal shape. Truss elements of the compound t-cross-sec., 2-formed equal parts	45.0	Steel Vst3sp5

116

Sec.8.3. Analysis of Defects and Damage of Bearing Building Structures

Analysis of defects and damage on bearing building structures made of eleven main buildings of enterprises of power, geographically located in different regions of the Russian Federation.

During the operation of building its reliability reduces over time. There is a need to repair. In this connection, to ensure the reliability of structures their maintainability plays the role, representing the adaptation of structures to the periodic inspection and maintenance [8.4].

There is a gradual depreciation of buildings and structures with time in the process of exploitation from occurring in them some of various defects and damage.

The survey of structural elements of the main buildings is the detection of defects, injuries and causes of them. The analysis of the revealed defects and damages will allow tracking damage of load-bearing elements of the skeletons of the main buildings depending on the duration of their operation.

It was found in according to the results of the survey that all main buildings had defects and damages for the various bearing structures.

General classification of damage to structures is made on the following grounds: the causes and nature of the damage, the type of impacts, the level of knowledge, the form and manifestations of the stage of accumulation. Damages, which were aroused by visual observation, were considered. Damage is considered as external or latent manifestation of irreversible processes of wear and ageing [8.4].

Fig.8.1 shows a diagram of damage of similar structures on the main buildings, expressed as a percentage, equal to the ratio of the number of similar structures with defects and damage to the total number of damaged structures for each of the objects.

$$U = \frac{\Delta\gamma}{Y}100\%$$

Where U is the damageability of similar structures; $\Delta\gamma$ - the number of similar structures with defects and damages on one object; Y - the total number of damaged structures on one object.

It was stated as a result of the generalization of the data of the survey of buildings that a substantial part of the structures is operated with injuries, the most dangerous of which is the destruction of concrete with denudation and corrosion of armature, corrosion of metal structures, and general deflections on the elements of the trusses.

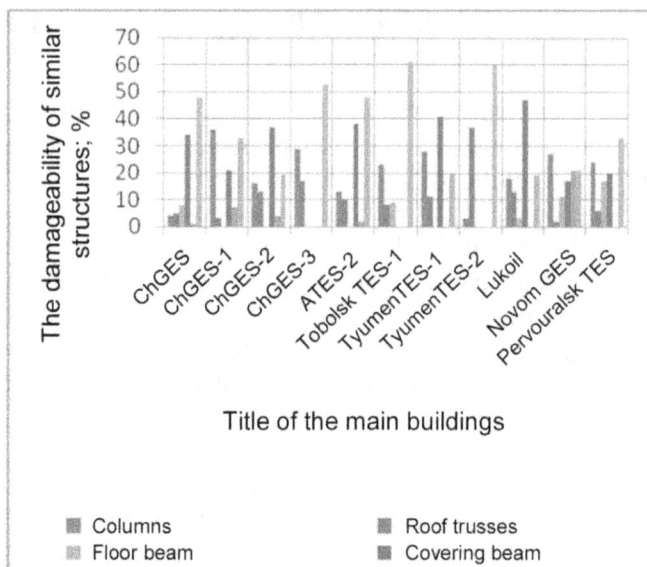

The damageability of similar structures; %

Columns
Floor beam
Roof trusses
Covering beam

Title of the main buildings

Fig.8.1. Diagram of a technical condition of the basic bearing structures of the main buildings

Diagram at Fig.8.2. and 8.3 demonstrates typical defects and damage of roof trusses and the structures of monolithic working platform.

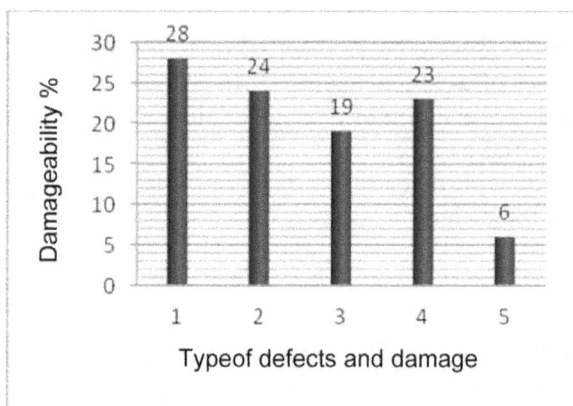

Damageability %

Typeof defects and damage

Fig.8.2. Diagram of the typical defects and damage on roof trusses: 1 – local deflections of the elements of the trusses; 2 – general deflections of the elements of the trusses; 3 – concentration of the technological dust on covering structures; 4 – destruction of the protective anticorrosive coating, surface corrosion; 5 – covering structures supports outside on joints at upper belt of roof truss

118

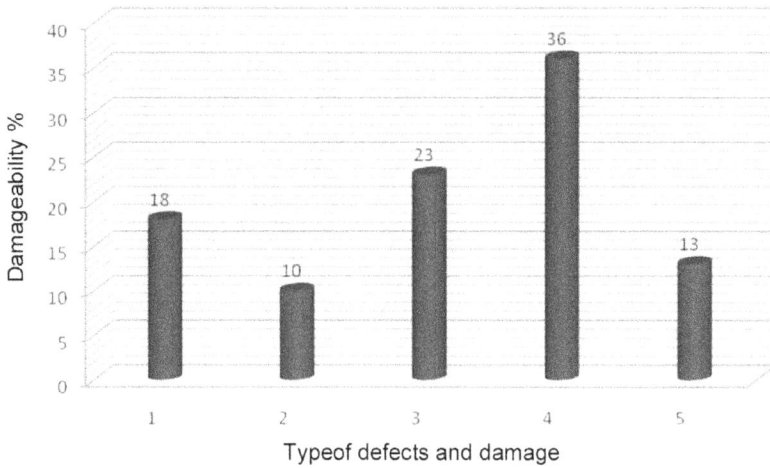

Fig.8.3. Diagram of typical defects and damage in the structures of monolithic working platforms: 1 – chips of the protective layer of concrete with denudation and corrosion of armature on the main floor beams; 2 – suspension of the technology pipelines to the longitudinal reinforcement of the main floor beams; 3 – chips of the protective layer of concrete with denudation and corrosion of armature on secondary beams; 4 – destruction of the protective layer of concrete in the shelves of the plates; 5 – crosscutting holes in the shelves of plates with denudation and corrosion of armature

The ratio of the number of the type of defect to the total number of defects by homogeneous structures in the surveyed objects was taken as the relative frequency of occurrence of defects and was expressed as a percentage.

$$\alpha = \frac{\Delta n}{N}100\%,$$

where α – the relative frequency of occurrence of the defect;
Δn – is the number of defects and damage of this type;
N – total number of defects on the same structures.

Sec.8.4. Reasons of Occurrence of Defects and Damage

The main causes of defects and damages of structures of the main buildings of energy enterprises, established with the performed surveys are presented in table 8.3.

Table 8.3

The reasons of occurrence of defects and damage

Title of structure	Material	Basic defects and damage	Reason of occurence
Columns	Reinforced concrete	The destruction of the protective layer of concrete with denudation and corrosion of working reinforcement	Corrosion of armature in the body of the concrete as a result of carbonization of the protective layer of concrete
		Corrosion of working armature. Cracks along armature	Impact of the aggressive environment
	Metal	Corrosion of working armature. Cracks along the valve Impact of the aggressive environment The destruction of the protective anticorrosive coating, surface corrosion places up to 50% of the surface area	Soaking of structures as a result of a malfunction of the systems of the technological pipeline
		The sandwich and crevice corrosion of structure	Violation of rules of operation
		Through holes in the walls of the columns	
Roof trusses	Metal	Local deflections of the elements of the trusses	Mechanical impacts during installation or operation
		General deflections of the elements of the trusses	
		Accumulation of technological dust on structures	Violation of rules of operation
Roof trusses	Metal	Soak. The destruction of the protective anticorrosive coat total surface corrosion on the elements of the truss	Percolation of roof's covering
		Girder supports outside of joints at upper belt of roof truss	Defect of installation
Covering beams	Reinforced concrete	The destruction of the protective layer of concrete with denudation and corrosion of the reinforcement. Chips of the protective layer of concrete	Corrosion of reinforcement in the result of violation of the protective layer of concrete.

Title of structure	Material	Basic defects and damage	Reason of occurence
Crane girders	Metal	Soak the beams of the covering. Corrosion of armature in the body of the concrete	Soak structures from leaks of roofing
		The total surface corrosion of covering metal beams up to 10% of the thickness of the cross-section	Soak structures as a result of the roof leaking
	Reinforced concrete	Delaminating of concrete protective layer over the entire length of the beam. Cracks on protective layer	Corrosion of reinforcement in the result of violation of the protective layer of concrete from the action of aggressive media
		Vertical cracks in the middle of the span crane beam width of disclosure 0.4 mm	Overload of structures. Reduction of concrete strength and decrease of the cross section of the armature as a result of corrosion
	Metal	The destruction of the protective anticorrosive coating on the beams. Surface corrosion on the elements of the beams	Soak structures as a result of the influence of the aggressive environment
		The lack of bolts in connection of crane beams	Violation of rules of operation
Main floor beams due to the working platforms	Reinforced concrete	Chips and cracks on the protective layer of concrete with denudation and corrosion of reinforcement of the main beams of the floor	Corrosion of armature in the body of the concrete as a result of violation of the protective layer of concrete from the action of aggressive media
		Suspension of technological pipelines to the longitudinal reinforcement of the main beams of the floor	Incorrect operation

Title of structure	Material	Basic defects and damage	Reason of occurence
Secondary beams and moonlit slabs of the working platforms	Reinforced concrete	Chips and cracks on the protective layer of concrete with denudation and corrosion of reinforcement of secondary beams of the floors	Corrosion of armature in the body of the concrete as a result of violation of the protective layer of concrete from the action of aggressive media
		The destruction of the protective layer of concrete with denudation and corrosion of reinforcement in the shelf plate monolithic floor	
		Through holes with denudation and corrosion of reinforcement on the shelf plate of monolithic floor	Incorrect operation
Covering slabs	Reinforced concrete	Cracks in the longitudinal ribs of roof slabs Corrosion of armature in the body of the concrete	Corrosion of reinforcement in the result of violation of the protective layer of concrete. Soak structures from leaks of roofing. Incorrect operation
		The destruction of concrete with denudation and corrosion of reinforcement on the shelf of the plate. Soak the cover plates	

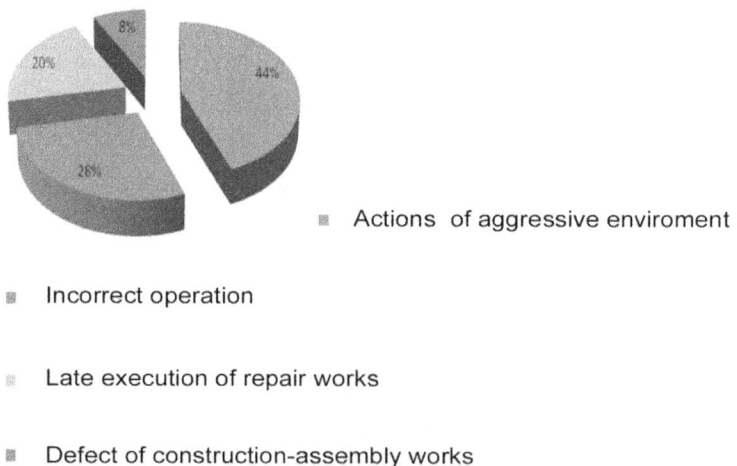

8%

20%

44%

28%

▦ Actions of aggressive enviroment

▦ Incorrect operation

▦ Late execution of repair works

▦ Defect of construction-assembly works

Fig.8.4. Diagram analysis of the causes of defects and damage due to the bearing structures of the main buildings

It was found that in most cases the damages are caused by influence of the external environment, namely aggressiveness, water constructions, mechanical damages during the operation.

The analysis of change of main building bearing structures technical condition depends on the period of operation.

There is a gradual depreciation of buildings and structures with time from arising in the process of exploitation some various defects and damage. A significant part of the defects and damage is formed and accumulated as a result of incorrect operation. Recording increases of damage of load-bearing structures during its operation and planning of repair-rehabilitation works will allow to consider reducing the likelihood of accidents and to increase safety of the main buildings of the enterprises of the energy sector.

The analysis of the revealed defects and damages will allow to track damage and to evaluate the change of a technical condition of bearing elements of the skeletons of the main buildings depending on the duration of their operation. According to the results of the analysis of the growth damage on the main bearing structures of the main buildings of enterprises of power depending on the period of operation, diagram presents a technical condition of bearing structures of the main buildings. Fig.8.5–8.10 presented a chart showing the change of a technical condition of the basic bearing structures of the main buildings of enterprises of power with the passage of time.

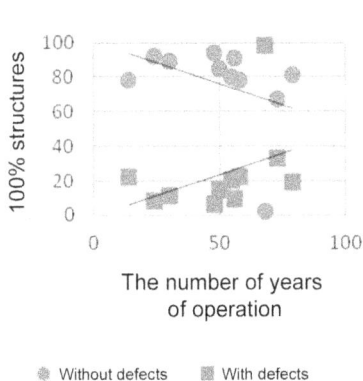

Fig.8.5. Diagram of change of technical status of columns

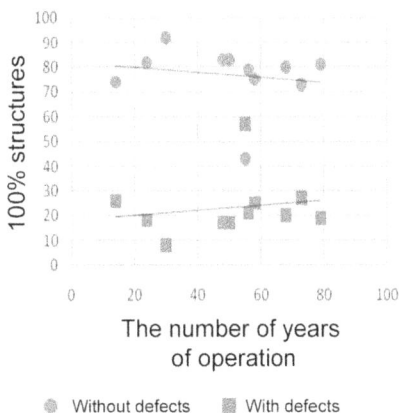

Fig.8.6. Diagram of change of technical status of roof trusses

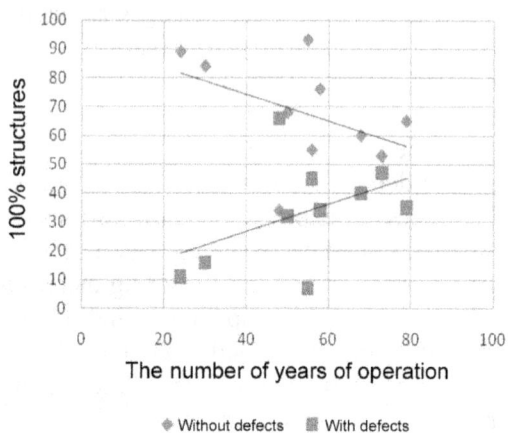

Fig.8.7. Diagram of change of technical status of floor slabs

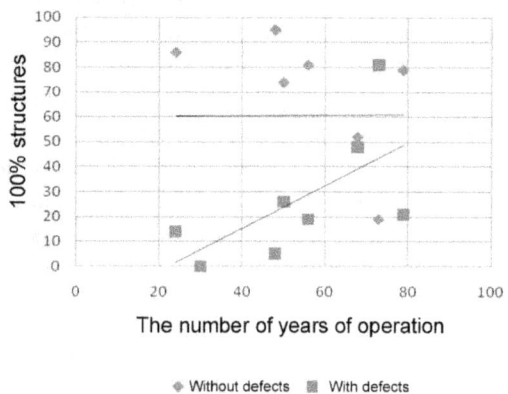

Fig.8.8. Diagram of change of technical status of crane girders

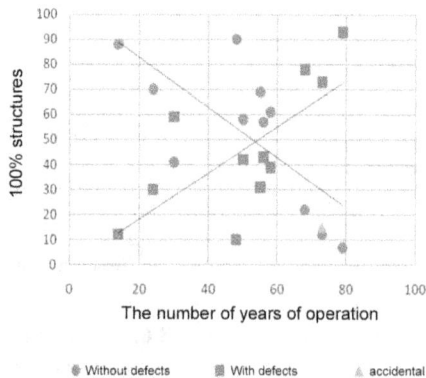

Fig.8.9. Diagram of change of technical status of working platforms

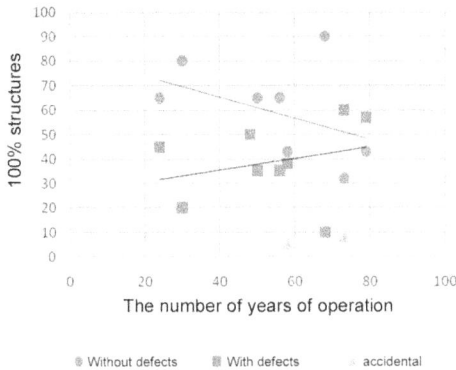

Fig.8.10. Diagram of change of technical status of covering slabs

The diagrams clearly shows, that with growth of the life of the building there is a reduction of the number of faultless structures and increase in the number of structures, which in the course of the survey were identified defects or damage.

Analyzing the presented diagrams of a technical condition of building structures it can be traced, that the fastest of all defects and damages are accumulated in the structures of a working platform, that is, these structures are the most damage in the main buildings. In connection with this, according to the working platforms a more detailed analysis carried out of the development of the most frequent types of defects and damage.

Sec.8.5. Study of Structural Wear of Working Platforms

Working platforms in the main buildings can be divided into the following types on a structural solution:

- beam type-monolithic;
- beam type with metal main and secondary beams and monolithic floor;
- beam type with metal main and secondary beams and fabricated reinforced concrete floor.

In the process of analysis of the working platforms identified the most common defects and damages. Diagram of typical defects and damage of the structure of monolithic working platform is shown in Fig.8.3.

The diagram for determination the speed of development of the most common defects due to dynamics of development of defects in the working platforms, presented in Fig.8.11.

The graphs shows that the more progressive nature of the development of the defects and damage take place in the structure of monolithic

125

reinforced concrete working platform. The destruction of the protective layer of concrete on individual elements in 50 years of operation of the building reaches 45%, and corrosion of reinforcement - 30%. After 50 years of operation the dynamics of defects in the working platforms increases dramatically.

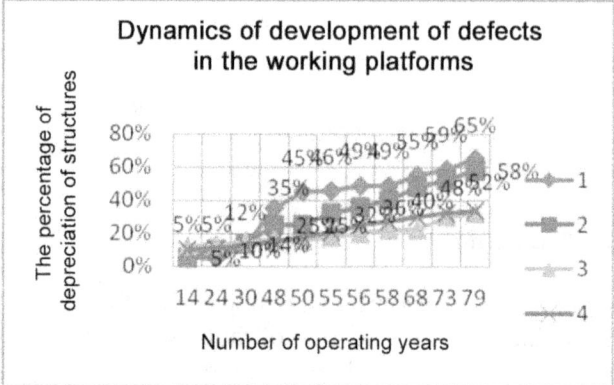

Fig.8.11. Dynamics of development of the defects in the working platforms of the main buildings of the enterprises of power engineering: 1 - the destruction of the protective layer of concrete with denudation of the armature (for 100% accepted by the square of one element); 2 - corrosion of reinforcement for reinforced concrete elements; 3 - corrosion of metal of the main beams floor; 4 - corrosion of metal secondary beams floor

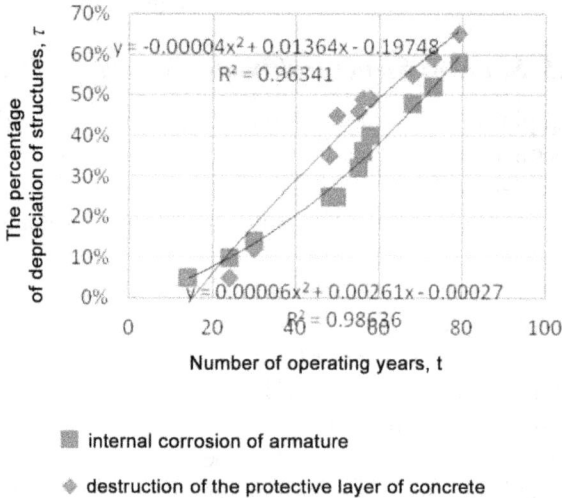

Fig.8.12. Dynamics of growth of the damage on the elements of the reinforced concrete platforms

Based on the results of the analysis, the rate of corrosion of reinforcement steel in reinforcement elements of the working platforms of the main buildings varies from 1.35 to 2,79 mm/year. That is, the data structures are operated under intense exposure environment, which causes a very high rate of corrosion.

Having studied the damage of a working platform, and having analyzed the growth rate of the corrosion of the armature in reinforcement elements of a working platform we built graphics and obtained the equations of approximation, allowing to determine the status of a protective layer of concrete and reinforcement of the exploited elements of reinforced concrete working platforms at any time.

There is an equation of an approximation for the determination of the dynamics of development of the destruction of the protective layer of concrete with denudation of the reinforcement steel

$$\tau = -0.00004 \ t^2 + 0.01364 \ t - 0.19748.$$

Equation approximation to determine the speed of development of corrosion of reinforcement steel in reinforcement element is presents as following

$$\tau = 0,00006 \ t^2 + 0,00261 \ t + 0,00027,$$

Where τ - the percentage of depreciation of the structure;

t – the number of years of operation.

Thus, it is established now on the results of analysis of design solutions of the main buildings that for placing of new technological equipment every year to these objects there are more new requirements for overall dimensions of the building in a plan and by altitude of spans. Study of the regularities of physical deterioration of elements of the skeletons of the main buildings of the thermal power stations showed that the fastest of all defects and damages are accumulated in the construction structures of a working platform, that is, the data structures are the most damaged in the main buildings.

The equations of approximation, received by results of the analysis of defects and damage, allow you to perform calculations on strength of reinforced concrete elements of the platforms on the faulty circuit with the view of reducing the cross-sectional area of reinforcement rods depending on the number of exploited years of the building.

Ch.9. Reliability of Corroding Structures

*V.D.Raizer**

Assessing the reliability of a structure deteriorating in time is a very important and challenging problem in design. The deterioration of structural material results in decreasing the load-bearing capacity in time and thus increasing the probability of failure. Irreversible change of structural material properties can be caused by corrosion in steels, decomposition in wood, ageing in polymers, as well as abrasion or erosion and accumulation of defects. Though the physics and mechanics of the material deterioration processes are different, their effects on material properties and structural reliability have some similarities that make it possible to apply more or less similar approaches. Therefore, the models and peculiarities of corrosion wear and its effects on reliability are discussed in this chapter.

Sec.9.1. Models of Corrosion Wear

Corrosion is an important factor in reducing the reliability and durability of different types of structures or equipment. From 10 to 12% of fabricated steel used in all structural industries is lost annually due to corrosion. In spite of widely used protection means, the quantity of steel destroyed is proportional to the accumulated supply of steel. Losses from corrosion are on the average between 2 and 4% of GDP in almost every country. About 30% of structural steel is subjected to atmospheric corrosion, and 75% is subjected to atmospheric and aggressive media corrosion simultaneously [Raizer, 1995]. Corrosion reduces the initial cross section of structural element and consequently their load-bearing capacity. Fig.9.1 shows the structural steel corrosion types.

Corrosion rate depends on aggressiveness of the corrosive medium and effectiveness of corrosion protection means. It can range from 0.05 mm/year to 1.6 mm/year. Damage to structural steel in soil depends on the exposure duration, as shown in Fig.9.2 for 16 types of soil. In sea water the corrosion rate of unprotected steel is so prohibitively high that virtually all marine structures are protected by more or less effective means. Similarly, damage to structural steel due to atmospheric corrosion is shown in Fig.9.3. Distribution of corrosion rate at the inner surface of fuel tanks along its height for is presented in Fig.9.4 for different types of fuel.

---

* Reliability Engineering Consulting, San Diego, USA

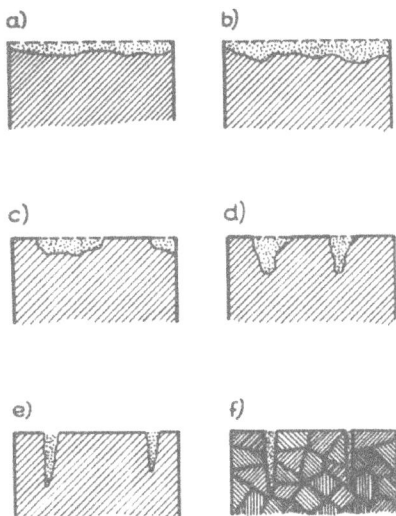

Fig.9.1. Types of structural steel corrosion:
a – uniformly distributed;
b – irregularly distributed;
c – corrosion spots;
d & e – pitted corrosion;
Types of structural steel corrosion:
a – uniformly distributed;
b – irregularly distributed;
c – corrosion spots;
d & e – pitted corrosion;
f – corrosion cracks.

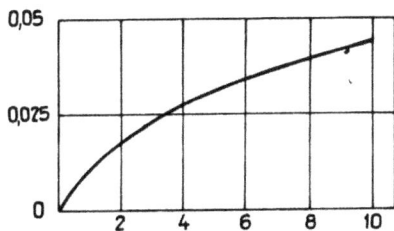

Fig.9.2. Corrosion of structural steel in soil

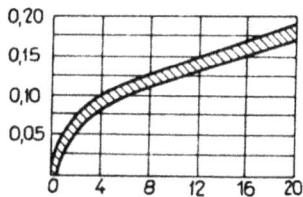

Fig.9.3. Corrosion of structural steel in open air

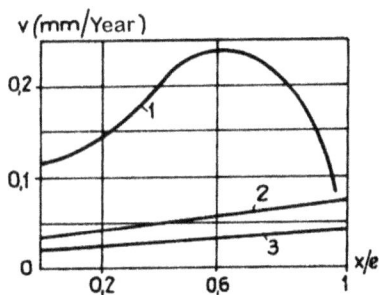

Fig.9.4. Variation of corrosion's speed: 1 – gasoline; 2 – kerosene; 3 – diesel

129

The evaluation of structural durability depends essentially on the type of model simulating the effect of aggressive environment. When modeling corrosion processes, there are important damage characteristics to consider, such as depth of corrosion waste (δ) and corrosion rate ($v = d\delta/dt$). Classification of mathematical models of corrosion (based on empirical approach) is presented in Table 9.1 [9.1, 9.2 and 9.3]. The kinetics of corrosion process in different metals subjected to different aggressive media looks very similar, and this fact presents an opportunity for applying these models in design. Generally, wear processes can be described by time-dependent random functions. The type of process depends on steel contents, structure's fabrication, corrosion protection, maintenance conditions, and other factors. Models of long-term processes are presented as random time processes, but their random character is caused by time-independent random parameters. Such kind of random processes are called "deterministic random processes" [9.4].

Table 9.1

Corrosion models

#	Models of corrosion	Functional relationship
1	$\delta = v_0 t$	Linear
2	$v_t = kt^{-n}$	Power
3	$\delta = a + b\lg t$	Logarithmic
4	$\delta = \ln(kt)$	Logarithmic
5	$v_t = v_0 \exp(-\alpha t)$	Exponential
6	$v_t = mt^2 \exp(-t/\tau)$	Exponential
7	$\delta = \delta_0[1 - \exp(-t/\tau)]$	Exponential
8	$\delta = \dfrac{a}{1 + b\exp(-ct)}$	Exponential
9	$v_t = \dfrac{t}{at^2 + bt + c}$	Fractionally linear
10	$\delta = \dfrac{\delta_0 t}{1 + at}$	Fractionally linear

When all loads F_i are presented as independent random variables, probability of no failure during service life can be expressed as:

$$P(n) = P[R_1 > F_1, R_2 > F_2, ..., R_n > F_n] \qquad (9.1)$$

Where $R_1, R_2, ..., R_n$ – values of bearing capacity in considered time intervals. Let's 12 n assume $R_n = R_0 \varphi(n)$, then $n = t$ – term of maintenance in years; R_0 is the initial random value of bearing capacity; $\phi(n)$ is a monotonically decreasing nonnegative function (i=1, 2, 3, n) satisfying to the conditions: $\phi(0) = 1$; $\varphi(\infty) = 0$; $d\varphi / dt < 0$. What should also be mentioned are the additive property of function φ *(t)* and independence of wear process at the subsequent time interval t_i of the previous process value at time t_{i-1}, i.e. $\varphi(t_1)\varphi(t_2) = \varphi(t_1 + t_2)$. $F_1, F_2, ..., F_n$ – loads corresponding to considered time intervals

Sec.9.2. Uniformly Distributed Corrosion Wear

This problem is discussed using, as an example, a section of steel cylindrical pipeline subjected to inner pressure, temperature changes and corrosion. The inner pressure F and yield stress of steel R_y are random variables with given distributions. The corrosion process is considered deterministic. The limit state condition is taken in the form of $S_i \leq Ry$, where S_i is the intensiveness of stress in the cylindrical shell. In accordance with the Guber-Mises condition [9.5], the general case can be expressed as:

$$S_i = \frac{1}{\sqrt{2}}\sqrt{(S_1 - S_2)^2 + (S_2 - S_3)^2 + (S_1 - S_3)^2} . \qquad (9.2)$$

In discussed situation $S_2 = 0$, and the radial and the tangential stress esare:
$$S_1 = \frac{FD_i}{2h}, S_3 = \frac{FD_i}{4h} - \alpha E \Delta\theta.$$

Here F is the inner pressure, and its maximum value is random for some of the time intervals; D_i is the inner diameter of the pipe; α-parameter of linear extension; E is the modulus of elasticity; $\Delta\theta$ is a temperature change (difference between temperature of the pipeline in operation and during assembly).

The reliability condition is expressed as follows:

$$\frac{3F^2 D_i^2}{16h^2} + \alpha^2 E^2 \Delta\theta^2 \leq R_y^2 . \qquad (9.3)$$

As temperature change presents an uncertain value with unknown distribution, thermal stresses are given as some part of the yield stress.

$$\alpha E \Delta\theta = R_y \sin\chi \qquad (9.4)$$

131

χ is a value of angle in the given interval $[0, \pi/2]$. The condition (9.3) can be written as:

$$F \le \frac{4h}{\sqrt{3}D_i} R_y \cos\chi. \tag{9.5}$$

Corrosion wear causes a reduction of tube thickness as $h = h_0\phi(t)$, where h_0 is the initial thickness. In accordance with the Table 1 one can take:

$$\varphi(t) = \exp(-t/\tau). \tag{9.6}$$

From (9.6):

$$h = h_0 - \delta[1 - \exp(-t/\tau)]. \tag{9.7}$$

Where δ is the depth of corrosion bubble. It is assumed that the corrosion process within interval t_2 is independent of the preceding values within interval t_1, so that $\phi(0, t + t_1) = \phi(o, t) + \phi(t, t + t_1)$. It is assumed also that time t takes only discrete values: $t=n$, where n is in years or months. An assumption is made for pressure F
that statistic data belong to a certain period of time, a month, for example. Only the maximum value from all observations, are selected. If the time interval is largeas compared with correlation zone, the 2nd Fisher-Tippet distribution of maximum values can be used [9.6].

$$P(x) = \exp[-(x/\xi)^{-\eta}]. \tag{9.8}$$

If $v_F = s_F / \overline{F}, \overline{F}$ are the coefficient of variation and the mean value, respectively, then parameters ξ and η are determined solving two equations with Gamma functions.

$$1 + v_F^2 = \Gamma(a)\Gamma(b),$$
$$\xi = \overline{F}/\Gamma(a). \tag{9.9}$$

Gamma functions are:

$$\Gamma(a) = \int_0^\infty e^{-z} z^{a-1} dz,$$
$$\Gamma(b) = \int_0^\infty e^{-z} z^{b-1} dz. \tag{9.10}$$

The case when $b=0$ and $\eta=2$ is ignored.
For yield stress is represented by the Weibull distribution.

$$P(x) = 1 - \exp[-(x/\omega)^\mu]. \tag{9.11}$$

Parameter μ is expressed via coefficient of variation $v_R = s_R / \overline{R}$:

$$v_R = \frac{\sqrt{\Gamma(1+2\mu) - [\Gamma(1+1/\mu)]^2}}{\Gamma(1+1/\mu)}. \tag{9.12}$$

The obtained values v_R and μ are then used to determine scale parameter ω.

Taking into account (9.8) and (9.11) the reliability function is written in the form:

$$P(n) = -\int_0^\infty \exp[-(\frac{4h_0 x \cos\chi}{\sqrt{3}D_i \xi})^{-\eta}] \sum_{i=0}^{n-1} \phi^{-\eta}(t)d[\exp\{-(x/\omega)^\mu\}]. \tag{9.13}$$

Example. After processing the statistic data of pressure in pipelines and yield stress the following values of the distribution parameters were defined: $\xi = 73.5$; $\eta = 65$; $\omega = 42.5$; $\mu = 23.5$. Coefficients of variations are: $v_F = 0.0201$; $v_R = 0.0522$. Temperature stresses (9.4) show their essential effect on pipeline reliability. When $\chi = \pi/3$, P (n) is close to zero. Values of P (n) values for different n are presented in the Table 9.2.

Table 9.2.

Probabilities *(Pn)* for different periods *n*

τ	χ	Values of function P (n)						
		Time in years						
		1	5	10	15	20	25	30
100	0	0.9989	0.9989	0.9989	0.9989	0.9987	0.9962	0.9860
100	6	0.9989	0.0087	0.9968	0.9880	0.9590	0.8600	0.6000
100	4	0.9560	0.8500	0.5800	0.1800	-	-	-
120	0	0.9989	0.9989	0.9989	0.9989	0.9989	0.9975	0.9872
120	6	0.9989	0.9941.	0.9941	0.9750	0.9600	0.8990	0.8060
120	4	0.9560	0.8790	0.6870	0.3790	-	-	-
150	0	0.9989	0.9989	0.9989	0.9989	0.9988	0.9985	0.9900
150	6	0.9989	0.9988	0.9980	0.9900	0.9760	0.9570	0.3200
150	4	0.9989	0.8820	0.7500	0.5200	0.3800	-	-

From (9.13) one can separate a term (or a formula) constituting the corrosion effect on reliability:

$$\lambda = \left[\sum_{i=0}^{n-1} \phi^{-\eta}(i)/n\right]^{\frac{\mu}{\eta+\mu}}. \tag{9.14}$$

Where λ characterizes a decrease of reliability due to the development of corrosion.

Parameter τ in (9.6) and in Table 9.2 defines intensiveness of uniform corrosion. Physical meaning of this value consists in decreasing of initial wall thickness. An essential decrease is possible at high values of $\tau = 100...150$.

Results of many experiments and real observations demonstrate [9.1, 9.3] the influence of stresses in structures on corrosion rate, especially in stress concentration zones. Relationship between corrosion rate and stress level can be either linear or non-linear. Assuming the stress level as a function of corrosion penetration depth such as $\delta = \alpha t^{\beta} \exp(kS_i)$, and substituting it in the formula for tangent stress in cylindrical shell $S_1 = \dfrac{FD_i}{2h}$, then the failure condition can be written as:

$$\frac{FD_i}{2\left[h_o - \alpha t^{\beta} \exp(kS_i)\right]} > R_y.$$

(9.15)

After expanding the term $\exp(kS_i) \cong 1 + kS_i$, into series, expression (9.15) can be yield:

$$F < 2\left[h_o - \alpha t^{\beta}\left(1 + \sqrt{3}kRy / 2\right)\right] / D_i.$$

(9.16)

Here the expression in round constitutes the stress state effect on corrosion rate. Assuming that the inner pressure defined by (9.8) and yield stress defined by (9.11), have the same distributions, and expressing the process of corrosion as a function of a discrete argument, the reliability function can be written in the form:

$$P(n) =$$

$$= -\prod_{0}^{\infty} \prod_{i=1}^{n} \exp\left\langle -\frac{2\left[x\left(h_o - \alpha i^{\beta}\left(1 + \sqrt{3}xk / 2\right)\right)\right]}{D_i \xi}\right\rangle^{-\eta} d\exp\left[-\left(\frac{x}{\omega}\right)\right]^{\mu}.$$

(9.17)

Expression (9.17) allows one to evaluate the reliability of pipelines, subjected to continuous corrosion and to take into account the stress state effect on corrosion penetration depth or corrosion rate.

Sec. 9.3. Irregular Distribution of Corrosion Wear

A problem of structural durability and protection against local corrosion is acutely important for many industries. Local corrosion leads to local degradation visible on the metal surface in the form of rusty spots, ulcers, pits and cracks (Fig.9.1). The process of degradation is random. It can be described using the following assumptions:

• Events that affect the occurrence of corrosion cavities at unrelated time intervals are independent;

• Probability of corrosion cavity occurrence at an arbitrary time interval t is proportional to the extent of the time interval with proportionality factor μ.

• Occurrence probability of two or more events within an extremely small time interval is an infinitely small value of a higher order of magnitude than that for either event.

A simultaneous realization of all these assumptions takes place for the simplest flow of events also known as a uniform Poisson process. Such a process can be described by a system of differential equations:

$$\frac{dP_0}{dt} = \mu P_0 .$$ (9.18)

$$\frac{dP_n}{dt} = \mu(P_{n-1} - P_n) .$$

The initial conditions for this system of equations are:

$$P_n (t) = 1, \text{ when } n = 0;$$ (9.19)

$$P_n (t) = 0, \text{ when } n = 1,2.$$

There is only one solution for the system (9.18) and together with the conditions (9.19) it can be presented as the Poisson distribution:

$$P_n(t) = \frac{[\mu(t - t_0)^n}{n!\exp[-\mu(t - t_0)]} .$$ (9.20)

Expression (9.20) is the Poisson distribution and constitutes the probability that at the $t \geq t_0$ instant the system is in the state of $n(n=1, 2, 3,...)$. If the number of cavities occurring within a time interval can be simulated by the Poisson distribution, the time of the next cavity occurrence follows the exponential distribution [9.7].

$$P(t) = \exp(-\mu t).$$ (9.21)

There is a rather limited volume of experimental data associated with studying the kinetics of pitting cavity formation and growth in structures, as well as on spread in their number. Experimental dependences were received in [9.8]:

$$\mu = \mu_{gr}(1 - e^{-\beta t}) .$$ (9.22)

Where μ_{gr}, β are empiric coefficients. The value μ_{gr} measured as the number of defects per a unit of structural surface, varies within a wide range. The most important parameters of irregular pitting corrosion include the maximum depth of a cavity, its diameter, and its area. The random value of cavity depth, δ_k (k – a random point on structural surface) is distributed in the final interval $[0, h_0]$, where h_0 is the structural element's thickness distributed uniformly, i.e.

$$P_\delta(x) = \begin{cases} 0 & at\ x < 0 \\ x/h_0 & at\ 0 \leq x \leq h_0 \\ 1 & at\ x > h_0 \end{cases} \qquad (9.23)$$

Distribution of the maximum depth for n cavities, i.e. $\delta_n = \max \{x_1, x_2, x_3,...,x_n\}$ is well known from theory of extreme values [9.9] and can be taken as exponential.

$$P_{\delta n} = \begin{cases} \exp[-n(h_0 - x)] & 0 \leq x \leq h_0 \\ 1 & x > h_0 \end{cases} \qquad (9.24)$$

The next important parameter is an across size of cavity or – assuming that it is of a circular shape – diameter (Fig.9.5). Let the depth of cavity is equal to x. Then the possible range of diameter variation is the chord AB whose length is $2\sqrt{2rx - x^2}$, and r is the external radius. An assumption is taken that the random value of diameter y_i has a uniform distribution within the interval $[0, 2\sqrt{2rx - x^2}]$.

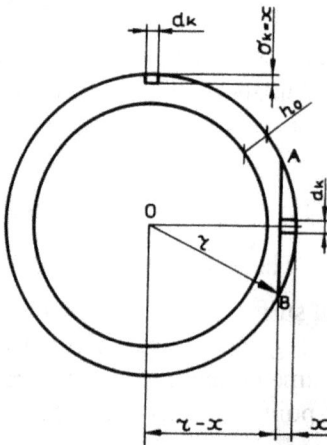

Fig.9.5. Cross-section of the pipeline with idealized cavity

136

$$P_d(y) = \begin{cases} 0 & \text{if } y < 0 \\ \dfrac{y}{2\sqrt{2rx - x^2}} & \text{if } 0 < y < 2\sqrt{2rx - x^2} \\ 1 & \text{if } y > 2\sqrt{2rx - x^2} \end{cases} \qquad (9.25)$$

Distribution of the maximum diameter for n cavities $d_n = \max(y_1, y_2, y_3, \ldots y_n)$ is:

$$P_{dn} = \begin{cases} \exp[-n(2\sqrt{2rx - x^2}, & 0 \le y \le 2\sqrt{2rx - x^2} \\ 1 & y > 2\sqrt{2rx - x^2} \end{cases} \qquad (9.26)$$

Third parameter of cavity is its area A_k. Knowing the maximum area is important for solving the problem. There are some uncertainties about this in the theory of order statistics. The point is that the maximum value of δ_n doesn't necessarily correspond to the maximum value of d_n. Accepting this notion would lead to a more conservative solution (i.e. the no-failure probability based on this notion would be less than the real one) – see case 1 below. If the notion is ignored, two kinds of versions can be offered: (a) – distribution of maximum depth δ_n and, depending on this, distribution of diameter d_k for one cavity in the first version (see case 2 below), and otherwise (b) – distribution of maximum diameter d_n, and, depending on this, distribution of depth δ_k in the second version (see case 3 below). Types of $PA(x)$ distributions are written for three cases:

Case 1:

$$P_{\delta n}(x) = \exp[-n(h_0 - x)], x \in [0, h_0],$$

$$P_{dn}(y) = \frac{y}{2\sqrt{2rx - x^2}}, 0 \le y \le 2\sqrt{2rx - x^2}. \qquad (9.27)$$

The area of the cavity A_k is equal to the area of the segment at Fig.9.5:

$$A_k = r^2 \arcsin \frac{y}{2r} - \frac{y}{2}\sqrt{r^2 - \frac{y^2}{4}} + [x - r + \sqrt{r^2 - \frac{y^2}{4}}]y. \qquad (9.28)$$

The maximum possible value of the cavity area A_k is when $x = h_0$ and $y = 2\sqrt{2rh_0 - h_0^2}$. In large diameter pipes x/r and $y/2r$ are very small numbers and the area can be approximated as $A_k = xy$, thus it follows:

$$P_{Ak}(A) = \frac{An}{2\sqrt{2r}} \int_0^{h_0} \exp[-n(h_0 - x)]x^{-\frac{3}{2}}dx. \qquad (9.29)$$

Where A_k is a random quantity uniformly distributed within $[0, A^*]$

Case 2:

$$P_{\delta n}(x) = \frac{x}{h_0}, x \in [0, h_0],$$

$$P_{dn}(y) = \exp\left[-n\left(2\sqrt{2rx - x^2} - y\right)\right], y \in \left[0, 2\sqrt{2rx - x^2}\right]. \qquad (9.30)$$

Distribution of A_k is:

$$P_{Ak}(A) = \frac{1}{h_0} \int_0^{h_0} \exp\left[-n\left(2\sqrt{2rx - x^2} - \frac{A}{x}\right)\right]dx. \qquad (9.31)$$

Case 3:

$$P_{\delta n}(x) = \exp\left[-n(h_0 - x)\right], x \in [0, h_0],$$

$$P_{dn}(y) = \exp\left[-n\left(2\sqrt{2rx - x^2} - y\right)\right], y \in \left[0.2\sqrt{2rx - x^2}\right]. \qquad (9.32)$$

It follows:

$$P_{Ak}(A) = \int_0^{h_0} \exp\left(-2\sqrt{2rx - x^2} - \frac{A}{x}\right)d\left[\exp\left(-n(h_0 - x)\right)\right]. \qquad (9.33)$$

The last case, as mentioned above, yields a conservative solution.

Example 1. *Reliability of a pipeline subjected to one-sided irregular corrosion.*

Dimensions of the corrosion cavity depth and diameter gradually grow so that a failure of pipe will occur when a cavity would grow through the wall. Time, t_n before this occurs can be calculated using the expression:

$$\int_0^{t_n} v(t)dt = h_0 - \delta_n. \qquad (9.34)$$

Where δ_n – maximum depth among n cavities; $v(t) = v_0 \exp(-\alpha t)$ – corrosion rate (Table 9.1,5). From (9.34) we get:

$$t_n = \frac{1}{\alpha}\ln\frac{v_0}{h_0 - \delta_n}. \qquad (9.35)$$

Time distribution $P(t_n < t)$ for a cavity to break through the wall can be expressed as:

$$P_n(t) =$$

$$= P\left\{\delta_n \geq \left[h_0 - \frac{v_0}{\alpha}\left(1 - \exp(-\alpha t)\right)\right]\right\} = 1 - \exp\left[-n\frac{v_0}{\alpha}\left(1 - \exp(-\alpha t)\right)\right]. \quad (9.36)$$

After averaging for n cavities, it follows:

$$P(t) = \sum_{n=0}^{\infty} \frac{(\mu t)^n}{n!} \exp\left\{1 - \exp\left[1 - \exp(-\alpha t)\right]\right\}. \quad (9.37)$$

Example 2. *Design of structural members under axial tension.*

A cylindrical element having a ring cross-section is considered. This element is subjected to irregular corrosion under deterministic load F. Denoting A_0 as the initial value of cross-section (t = 0), and A_k as the area of cavity with given distribution $PA_k(A,)$ allow one express of no failure as follows:

$$F / (A_o - A_k) < R_y \quad or \quad A_k < A_0 - F / R_y. \quad (9.38)$$

Substituting the last expression into distribution function as an argument and averaging with respect to n and R_y, the probability of no failure at instant t is:

$$P(t) = \exp(-\mu t) \sum_{n=0}^{\infty} \frac{(\mu t)^n}{n!} \int_0^{\infty} P_{Ak}\left(A_0 - \frac{F}{R_y}\right) p\left(R_y\right) dR_y. \quad (9.39)$$

Where $p(R_y)$ – is probability density of yield stress. The following data are taken in the numerical example: external diameter D = 6.26in; initial thickness h_0 = 0.24in; F = 58.08t; $\mu = \mu_{gr}[1 - \exp(-\beta t)]$ and β = 0.05; $\overline{R}_y$ = 290 Mpa; s_{Ry} = 25 Mpa. Parameters of the cavity are $\overline{d}_k = \overline{\delta}_k = 0.008$ in. Results of numerical calculations are plotted at Fig.9.6, 9.7.

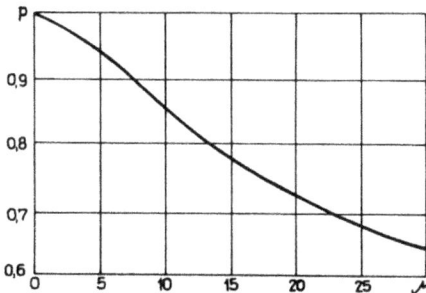

Fig.9.6. Reliability function for Example 1

139

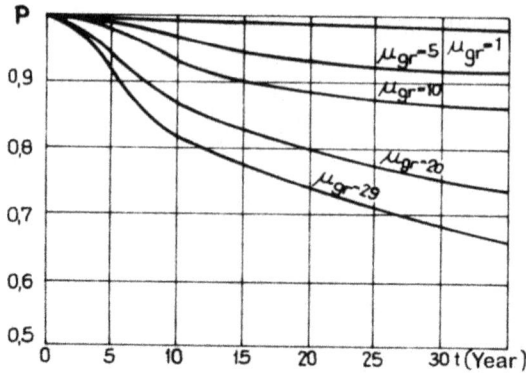

Fig.9.7. Reliability function for Example 2, at different numbers of cavities

Sec.9.4. Calibration of Partial Factor

The partial factor to model uncertainties can be determined comparing the model with similar structures operating in normal or aggressive environment. Let us consider the structure under load F and with resistance R. When random value of the load maximum for a definite period of time (one year, for example) has distribution $P_F(x)$ and the annual load maximums are independent random values, the reliability function can be written as

$$P_1(n) = \int_0^\infty P_F^n(x)dP_R(x).$$
(9.40)

It is assumed that there is a structure operating in an aggressive environment and subjected to uniform corrosion. To provide sufficient reliability in the design, additional structural material is necessary for increasing the cross-sectional area. The condition of no failure is

$$\tilde{F} \le \gamma_D \tilde{R}.$$
(9.41)

Though the corrosion process is continuous in time, let us consider $\phi(t)$ as a function discrete variable n. Condition (9.41) for the nth year could be rewritten as

$$F \le \gamma_D R\phi(n).$$
(9.42)

Reliability function will be

$$P_2(n) = \int_0^\infty \prod_{i=1}^n P_F\big[\gamma_D x\phi(i)\big]dP_R(x).$$
(9.43)

140

The equation for defining γ_D can be derived from equations (9.40) to (9.43).

$$\int_0^\infty \prod_{i=1}^n P_F\left[\gamma_D x\phi(i)\right]dP_R(x) = \int_0^\infty P_F^n(x)dP_R(x). \qquad (9.44)$$

For a non-corrosive structural element subjected to axial force (9.41) can be presented as

$$F \le RA_0. \qquad (9.45)$$

Where A_0 – initial cross-section area. Function of reliability will be

$$P_1(n) = \int_0^\infty P_F(xA_0)dP_R(x). \qquad (9.46)$$

For corroding structural element, the cross-sectional area is $A_0\gamma_D$, and $\gamma_D > 1$. The failure condition can be expressed as

$$F \le \gamma_D A_0 \phi(n) R. \qquad (9.47)$$

Reliability function (9.43) will be

$$P_2(n) = \int_0^\infty \prod_{i=1}^n P_F\left[\gamma_D A_0 x\phi(i)\right]dP_R(x). \qquad (9.48)$$

Equality (9.44) for the fixed value of n allows one to determine γ_D. The Fisher-Tippet distribution (9.8) was chosen for $P_F(x)$. Equality (9.44) will be yeild

$$\int_0^\infty \exp\left[-n\left(\frac{xA_0}{\xi}\right)^{-\eta}\right]dP_R(x) = \int_0^\infty \prod_{i=1}^n \exp\left[-\left(\frac{\gamma_D A_0 x\phi(i)}{\xi}\right)^{-\eta}\right]dP_R(x). \qquad (9.49)$$

Then it follows

$$\gamma_D = \frac{1}{[n\sum_{i=1}^n \phi^{-\eta}(i)]^{\frac{1}{\eta}}}. \qquad (9.50)$$

Introducing the corrosion model in the form of (9.6) and expanding the sum in (9.50) in series we can get

$$\gamma_D = \frac{1}{\left\{n\left[\exp\left(\frac{\eta}{\tau}\right) + \exp\left(\frac{2\eta}{\tau}\right) + ... + \exp\left(\frac{n\eta}{\tau}\right)\right]\right\}^{\frac{1}{\eta}}}. \qquad (9.51)$$

After transformation we get

$$\gamma_D = \frac{\exp\left(\dfrac{n+1}{\tau}\right)-1}{n\left[\exp\left(\dfrac{\eta}{\tau}\right)-1\right]} \,. \tag{9.52}$$

The modal factors obtained in accordance with (9.52) are given in Table 9.3.

Table 9.3

Partial model factor in corrosive medium

n	η	γ_D In heavily aggressive medium $\tau = 100$	γ_D In moderately aggressive medium $\tau = 150$	γ_D In mildly aggressive mtdium $\tau = 200$
	10	1.0666	1.0461	1.0364
10	20	1.1304	1.0828	1.0612
	30	1.2098	1.1280	1.0920
	10	1.0657	1.0439	1.0337
15	20	1.1374	1.0851	1.0618
	30	1.2263	1.1354	1.0958
	10	1.0665	1.0433	1.0327
20	20	1.1443	1.0881	1.0632
	30	1.2401	1.1424	1.0999

Aggressiveness of environment can be classified depending on parameter τ: $\eta =100$ for heavily aggressive, $\eta = 150$ for moderately aggressive, and $\eta =200$ for mildly aggressive corrosion media.

References

Ch.5

5.1 Efremova S.V, Stafeev K.G., Petrochemical research methods of rocks, Manual. M., Nedra, Publ. House. (Петрохимические методы исследования горных пород).

5.2 Grunin I.Yu., Budko V.B. (2009), Scientific-methodological principles of visual and measuring control in the building examination, Textbook , Under ed. Troitsky-Markov T.E., M., VEMO, 166 p. (Научно-методические принципы визуально-измерительного контроля в строительной экспертизе).

5.3Grunin I.Yu , Budko V.B, Lipin D.A., Gorkin D.S. Belich,Yu.V., Blinova Yu.M., (2011), Application of the VIK, organoleptic and laboratory research methods in carrying out express-diagnostics of objects in the field, Training manual, Troitsky-Markov T.E M., 320 p.(Применение ВИК, органолептических и лабораторных методов исследований при проведении экспресс-диагностики объектов в полевых условиях)

5.4. http://prosvetlenie.net/show_content.php?id=82

5.5. SNiP 22-01-95. Geophysics of hazardous natural phenomena. (СНиП 22-01-95. Геофизика опасных природных воздействий).

5.6. Instructions for engineering Geology and Geoecology interaction all over investigations in the Moscow city (2004), M., Moskomarkhitectura. (Инструкция по инженерно-геологическим и геоэкологическим изысканиям в г. Москве).

5.7. Vasilyev A.G., Kopeikin V.V., Morozov P.A. (2002), Ground penetrating radar in underwater archaeological studies IZMIRAN, Troitsk. Fund of underwater archaeological research. E. Blavatskogo, Ancient Bospor. №5. (Георадар в подводных археологических исследованиях).

Ch.6

Ch.7

7.1. DBN B.1.1-12:2006.Construction in seismic regions of Ukraine / Ministry of construction of Ukraine. – Kiev, Ukrarchstroiynaform, 2006, 84p. (Строительство в сейсмических районах Украины)

7.2. DBN B.1.1-5-2000. Buildings and structures at the undermining territories and subsiding soils. H. II: Buildings and constructions on subsiding soils / State Committee for construction, architecture and housing policy of Ukraine. – Kiev, State construction Committee of Ukraine, 2000, 84p. (Здания и сооружения на подрабатываемых территориях и просадочных грунтах. Ч. II: Здания и сооружения на просадочных грунтах).

7.3. ISO 4866:2010. Mechanical vibration and shock. Vibration of fixed structures, Guidelines for the measurement of vibrations and evaluation of their effects on structures, 48p.

Ch.8

8.1.Kuznetsov I.P., Ioffe Yu.R... (1985), Design and construction of thermal power plants, 3 ed., added. M., Energoatomizdat Publ.House. (Проектирование и строительство тепловых электростанций).

143

8.2.Dobromislov A.N. (2008), Diagnostics of damages of buildings and engineering structures, M., MGSU. (Диагностика повреждений зданий и инженерных сооружений).

8.3. The prevention of accidents of buildings and constructions: Collection of scientific works (2009), under ed. Eremin K.I., №8. M. (Предотвращение аварий зданий и сооружений).

8.4. SP 13-102-2003. The rules of examination of bearing building structures of buildings and structures, Electronic reference book of standardizing documents "System Info-EXPERT (Moscow)". (Правила обследования несущих строительных конструкций зданий и сооружений).

Ch.9

9.1 Raizer V. (2009), Reliability of Structures. Analysis and Applications, Backbone Publ. Co., USA, 145p. (Надежность конструкций. Анализ и приложения).

9.2 Ovtchinnikov I.G. (1982), Mathematical Corrosion Prognosis of Structural Steel Elements, in VINITI typescript No2061, Moscow (in Russian), 15pp. (О математическом прогнозировании коррозии металлических элементов конструкций).

9.3 Tsikerman L.Ya. (1977), Computerized Diagnostics of Pipeline's Corrosion, Nedra, Publ. House, Moscow (in Russian), 319pp. (Диагностика коррозии трубопроводов с применением ЭВМ).

9.4. Middlton D. (1961), Introduction to a Statistical Communication Theory, Book 1, Soviet Radio Publ. House Moscow (in Russian), 782pp.(Введение в статистическую теорию связи).

9.5 Lellep Ya.A.,Khannus S.Kh. (1987), Large Deflections of a Cylindrical Shell from Rigidity Plastic Material, International Applied Mechanics, Publ. Springer NY, Vol.23, No 5, 437-442pp.

9.6 Gumbel EJ (1967), Statistics of Extremes, Columbia University Press, New York. (Статистика экстремальных значений).

9.7. Kapur K.S. and Lamberson L.R. (1977), Reliability in Engineering Design, Wiley, New York., 604pp. (Надежность и проектирование систем).

9.8. Zaslavski I.N., Flaks V.Ya.,Chernyavski V.L.,(1979), Durability of Buildings in Ferrous Metallurgy, Stroyizdat Publ. House, Moscow (in Russian), 72pp. (Долговечность зданий и сооружений предприятий черной металлургии).

9.9. Galambos J. (1987), The Asymptotic Theory of Extreme Order Statistics. Robert Krieger Pub. Co., Malabar, Florida. (Асимптотическая теория экстремальных значений).

Part III. Analysis of Structural Safety and Reliability

Ch.10. Methods of Estimation of Residual Buildings Life

*Sushchev S.P., Adamenko I.A., Samolinov N.A.**

Sec.10.1. Definition of a Residual Resource

The task of estimation of a residual resource of a building is currently one of the most pressing tasks in the sphere of safety of buildings and structures operation, which requires a solution in order to implement forecasting in time of the value of that resource until exhaustion of the building consumer cost.

Consumers cost is considered here as a numerical expression of the quality of the building in the functional dependence from time to time. It can be defined by indicators of various kinds: the size of physic-mechanical characteristics of the building $R=f(R_0, t)$, the cost of a building $C=f(C_0, t)$ and the duration of life of the building $T=f(t_0, t)$ [10.1]. Accordingly, where life (i.e. the consumers cost of the building at this point of time, expressed in the performance of its quality) is determined by the dependence

$$K_i = f(P_{i,t}/P_{i,0}), \qquad (10.1)$$

where $i = 1, 2, 3$; $P_{i,t} = \{R, C, T\}$.

The values included in the expression (10.1) and determining the quality of the whole buildings and individual structures have a random character. This is because of the variation in time of their intrinsic properties (materials) and in the general case, the external conditions (loads and actions), as well as the statistical properties of original factors. Characteristics and performance of the last to the moment of finishing the construction of the building is determined by its initial reliability which is reduced from the first day of operation.

Conditions of this reduction will be gradual in a sufficiently stable operation and forecast of the moment of exhaustion of the residual resource can be refined to clarify the information regarding actual state of structures in the process due to implementation of the planned monitoring.

As a result of the impact of the accidental loads (seismic, blasts and mechanical effects, a sharp violation of the technological process, defor-

---

*Center for Research of Extreme Situations, Moscow.

mation of the undermined territories, soaking karsts and loess rocks) intensive wear of building structures occurs in a short time step like manner.

In these cases determination of the conditions of structures and properties of their materials should be made immediately after the action.

The task of definition of a residual resource can be solved with different levels of severity. The highest level corresponds with the condition of all kinds of uses of opportunities (and benefits) of the theory of probability according to the time characteristics of scattered values. At the relatively low level certain distribution is used, and any response may be represented only by its mean value and standard deviation. The lowest level corresponds to deterministic design.

And even in a deterministic setting the task of a residual resource of the building relates to such type of tasks, for which a preliminary physical analysis does not shows [10.2] the exact form of the formula and only allows a choice of formula (the dependence of the coefficient of safety «k» from time t) from a very wide range. Often the empirical formula of polynomials of varying degrees is being used. There is a problem of optimal degree of the polynomial. They are guided by the following considerations:

- polynomial of low degree gives a rough description of the empirical material, and a polynomial of high degree will not mitigate the noise of the experiment;

- selection of the optimum degree of the polynomial is based on the assumption that the functional relationship with a sufficient approximation can be represented by a polynomial of some degree n:

$$y = \sum_{j=1}^{n} a_j x_j$$ and the measured values y_k contain only random errors η_k;

- it is assumed that the errors are independent and follow the normal distribution law with the same variance σ^2 with equally accurate measurements, or with variances η_k/ω_κ when unequally accurate measurements (ω_k - known weights coefficients);

- it is possible and appropriate to solve on this basis the problem of the residual resource with the application of the theory, giving the answer to the question of when will appear the limit time and how likely it is to be expected. In this case at last when sufficient experimental material is gained, it is more promising in the sense of achieving the objective assessment of the value of a residual resource to use random functions.

Determination of the initial potential is for the time $t = t_0 = 0$ (i.e. for the moment of the beginning of exploitation of the building after its construction -major repair, or reconstruction) for the project with the involvement of the executive drawings. It boils down to the definition of the initial safety factor on the considered potential (the formula (10.1)). Dependence, reflecting the nature of decreasing functional qualities of the structure must be sufficiently justified. The procedure for selection of such dependence should take into account (in the absence or insufficiency of information material), at least logically, the process of losses in time of the design ability of structure to perform its functions. Otherwise, the extrapolation of the selected law of the behavior of a structure of the re-viewed parameter can lead to substantial errors, sometimes with severe consequences.

If we take into account the above considerations and consider the ex-isting or proposed development of methods of definition due to a residual resource of a building, it should be noted that:

– the vast majority of them is based on the deterministic view of the process of changing properties of a structure in time;

– in connection with a limited number of moments (in time), on which field measurements of structures characteristics are carried out, it shall be adopted as sufficient building their functional dependence in three stages. Such an approach predetermines dependence from the linear to quadratic and makes it necessary to specify additionally by the type of connection between variables (linear, quadratic) on the basis of logical reasons.

$$k = f(t^{\alpha}), \qquad (10.2)$$

when $\alpha = 1$, function is linear. When $\alpha \neq 1$ this relationship is non-linear. Indicator α can take the intermediate values. In [10.4] the technique of definition of a residual resource of industrial smoke (or ventilation) pipes is presented with a linear approximation of the changes over time in its VAT. This leads almost impossible to extrapolate the development of process on the interval from the time of the final examination till the time of limit state design with no fear to a significant extent away from the truth.

The assumption of the possibility of describing change of factor k in the law of the square of the parabola with the axis of symmetry axis of ordinates with the direction of the branches of the parabola in the direc-tion of the negative y-axis leads [10.3-10.5] to the expression

$$t_H = t \sqrt{\frac{k_0 - 1}{k_0 - k}} , \qquad (10.3)$$

where t is the time, when $k = 1$, and $k = k(t)$; $k_0 = k(t = 0)$.

We considered the possibility to use for description of the law of coefficient k change square parabola with a symmetry axis as abscissa axis, the vertex at the point $O(0;0)$, and the boughs of which are directed towards negative values of abscissa, i.e., $y^2 = -2px$, or $k^2 = 2p(t - a)$; and $a \le x \le 0$.

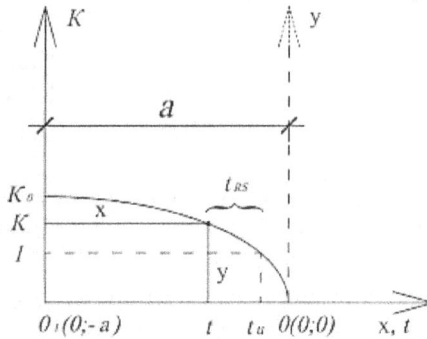

Fig.10.1

Where

$$p = (1 - k_0^2) / (2t_H); a = -(k_0^2 t_H) / (1 - k_0^2) \qquad (10.4)$$

The choice of this dependence can be explained, in our opinion, as having bigger fit (slow decline in the functional quality of the structure in the initial period of operation and rapid decline of its end-period) compared to (10.3) the law of variation of the values of k (t) in the interval from the time of the beginning of exploitation of structure until the moment of its limit state $(k = 1)$.

Thus, in essence considered deterministic approach to the design of structures [10.6] in two stages:

– stresses (deformations, displacements) are determined in structures that are exposed to external loads;

– the comparison is made with the calculated stresses (deformations, displacements) in the structure with their limit values.

For the considered task another stage is added: determination of the time of equilibrium between the calculated and limit values.

Sec.10.2. Application of Probabilistic Methods

The statistical interpretation of the safety factors opens the possibility for a reasonable way of assessing the reliability of the obtained results. Deterministic task turns into the problem of determination of probability of possible period of permitted operation of a building on the basis of probabilistic characteristics of random external conditions and random parameters of structures.

The use of probabilistic methods requires a sharp increase in the volume of information about external actions, as well as information about the materials of structures. The increase in the volume of the necessary information is a natural fee for a more accurate prediction of the behavior of structures and for greater authenticity of conclusions about its reliability and durability.

The main provisions of the probabilistic approach are:

– external conditions of exploitation of a building are random processes;

– principal indicator of the reliability is taken by the probability of parameters of the system being in a certain acceptable region, violation of normal operation leads to the exit from this area;

– failure of structure is, as a rule, is the result of the gradual accumulation of damage;

– assessment of the conformity of the real risk of an accident in accordance with requirements for design safety [10.9] is an integral part of the definition of a residual resource.

In accordance with (10.4) the residual resource of structures t_{RS} (the time it reaches the limit state from the moment of t):

$$t_{RS} = t_H - t = t_H \left[1 - \frac{k_0^2 - k^2}{k_0^2 - 1} \right]$$ (10.5)

The values included in the expression (10.5) are different on the basis of statistical certainty:

$$t_{RS} = f(t, k_0, k)$$ (10.6)

where t is the time argument, a deterministic variable the value of time;

t_{RS} - time of residual life is a random function of time;

k - random function of the time:

$$k = k \left[\varphi(R_1 t) / \psi(N) \right]$$ (10.7)

where $\varphi(R_t)$ is a random function the quality of structure in time;

$\psi(N)$ - non-random function of the loads on the structure in time (to be determined by codes);

k_0 is the random variable at time $t = t_0$.

$$k_0 = k \left[\varphi(R_1(t=0)) / \psi(N) \right]$$ (10.8)

i.e. it can be considered as a realization of the random function (10.7) for $t = t_0$.

Primary (at the moment of time $t = t_0$), the value of k_0 is a random variable. It is assumed that the distribution of the individual implementations k_{0j} corresponds to the normal law, determined by:

– the average value

$$M_{k0} = \frac{1}{n}\sum_{j=1}^{n} k_0 \qquad (10.9)$$

– as the standard deviation σ_{k0} it is in advance unknown it can be approximately measured by the empirical standard:

$$S_{k0} = \sqrt{\frac{1}{n-1}\sum_{j=1}^{n}\left(k_{0j} - M_{k0}\right)^2} \qquad (10.10)$$

Thus, the confidence interval, defining the boundaries of the feasible values of R_0 with the reliability of P is equal to

$$M_{k0} - \alpha_0 S_{k0} \leq k_0 \leq M_{k0} + \alpha_0 S_{k0} \qquad (10.11)$$

or

$$\varepsilon_{R_0} = \frac{\alpha_0 S_{k0}}{M_{k0}} \qquad (10.12)$$

here $\alpha_0 = f(P)$ is the quintile value in determining the P. In accordance with [10.2]

$$\alpha_0 = q(P,n)\frac{\sigma}{\sqrt{n}}$$

Empirical standard S_{R_0} defined by the formula (10.10) is used if there is an unknown in advance the value of the average quadratic error. Thus,

$$\alpha_0 = \frac{q(P,n)S_{k0}}{\sqrt{n}} \qquad (10.13)$$

The values of $q(P,n)$ depending on the specific values of P and n are accepted by the table [10.2].

Interpolation of the values $q(P,n)$ is allowed only on the argument of «n». At $16 < n < 60$ error of linear interpolation does not exceed 7×10^{-8} for $P = 0,999$, 4×10^{-3} for $P = 0.99$ and 2×10^{-3} for $P \leq$ of 0.98. At $60 < n < 100$ error of linear interpolation is not greater than 10^{-3}. The same kind of reasoning leads to expressions of the random variable k_t at the moment $t = t_i$. They will be identical to the expressions (10.9)-(10.13) with the replacement of the index «0» on the index «t_i». Function (10.5) if the random nature of the values of k_0 ($t = 0$) and k_t ($t = t_i$) is a function of random variables from non-random parameter t. Set of random realizations of k_0 and k_t values is interpreted geometrically as a random point with coordinates (k_0, k_{ti}) on the plane $[\ k_0\ O\ k_{ti}\ ]$ located in the specified rectangle. This problem was solved previously applied to the underground mine workings [10.8].

However, in this case the problem can be simplified. The value of «k_0» is defined according to the initial data, taken from the project (executive drawings) and is, in essence, a deterministic value. On this premise there is no statistical variation of the parameters of structures and their numerical characteristics, and the value of k_0 in the expression (10.5) can be adopted as deterministic. The function t_{RS} is a random function of the non-random argument t with additional signs of a function of

random variables $\quad x = R_{t^2} - \psi(N), y = \dfrac{1}{R_{O^2} - R_{t^2}}\quad$ with a mathematical

expectation

$$M[t_{RS}] = t\left\{(M[\phi(R_t^2)] + \psi^2(N))(M[\phi(R_0^2) - \phi(R_t^2)] + K_{xy}\right\}. \quad (10.14)$$

Here K_{xy} – correlation moment, which determines the degree of interdependence (close connection between the «x» and «y

$$K_{xy} = M[(x_i - m_x)\ (y_i - m_y)]. \quad (10.15)$$

When $K_{xy} \approx 0$, then «x» and «y» are the independent variables. Standard

$$S_{RS}^2 \approx \sigma_{RS}^2 = \sigma_x^2\sigma_y^2 + m_x^2\sigma_y^2 + m_y^2\sigma_x^2, \quad (10.16)$$

σ_x, σ_y, m_x, m_y are defined for the random variables on known [10.7] formulas at every stage (t_i) for identification the numerical values of the characteristics of structures

The confidence interval for t_{RS}

$$M[t_{RS}] - \alpha S_{RS} \leq M[t_{RS}] \leq M[t_{RS}] + \alpha S_{RS}. \quad (10.17)$$

Here α - quintile, defined by a table with a given number of trials and by the required level of reliability of the obtained results.

The above technique can be effectively used only under condition of sufficient adequacy of the arrays of the initial data. There is a need for a number of training specialists conducting the examination directly at the site as well as sufficient set of modern devices and instruments including methodical provision of the work is required. This goal is the methodical manual "Examination of buildings structures with the use of the upgraded mobile diagnostic complex for estimation of technical condition of build-

151

ings and structures (ММДС)", created by the group of specialists of LLC "TSIEKS" in the development of the SP 13-102-2003 to use when checking the technical condition of building structures. The paper summarized and used the most up-to-date regulatory documents and other technical resources. This paper takes also into account peculiarities of conducting of inspection of buildings, exposed to fire or earthquake forces. There were summarizing the main provisions regulating the general procedure of preparation, examination and registration of results of inspection of bearing building structures with the use of ММДС.

Sec.10.3. General Methodical Principals

The approach to the composition and content conforms to the essential methodological (basic) principles survey of buildings which include:

– representativeness of the primary data of full-scale survey. It is being achieved by logical place-wise and by number of structural plots for a detailed inspection, in order to ensure at the same time representativeness of the initial data with a minimum of labor.

– the veracity of primary data of the field survey (with a certain level of reliability), provided by statistical processing of the numerical results of full-scale survey with the exception of the preliminary gross errors;

- application of a unified regulatory framework as the main regulator of the quality of the survey;

– harmonizing methods for the inspection of structures and technologies of working with the devices of the complex equipment;

– the use of scientific methods of analysis and synthesis of a survey in the process of formulation of conclusions on the technical condition of building and the formulation of recommendations for the restoration of the functional quality of building structures.

Adherence to these principles will help to be close to the actual technical condition of building structures and in general to the whole building, avoid significant errors in the assessment of this state, identify effective measures for the restoration of the functional properties of structures.

The methodology sets out the procedure for conducting a survey of load-bearing structures of buildings. The body of works and the technology of their implementation at every stage are being determined, including methodologies for assessing the technical condition of structures, their actual carrying capacity, and the principles of making reasonable technical solutions for repair and reconstruction activities or methods of strengthening.

The importance of the above-mentioned principles will show on the example of the statistical processing of the results of measurements of the strength of the structural material with non-destructive methods.

As a result of measurements of the strength of a material of bearing structures by the devices of non-destructive control (included ММДS: ultrasonic device for determining the integrity and strength of concrete, bricks UK-15M; sclerometer electronic ОМШ-1Е; a device for determining the strength of concrete PIB) primary data on the strength of the material were received. To ensure the representativeness of the source data, the number of individual measurements shall be not less than 12–15.

To demonstrate the methods of statistical processing of measurement results let us consider the example of many of the initial data, obtained in a cross sounding columns of the building cross-section 250×500 mm by ultrasonic device UK -15M.

Measurement results and their statistical treatment are presented in table.10.1.

Table 10.1

Element	Area	№	t mks	B mm	V m/sec	R MPa	R_H MPa	R_H Correction. MPa	$m(R^H)$ MPa	$s(R^H)$ MPa	R_b MPa
1	2	3	4	5	6	7	8	9	10	11	12
Column basement	B/6	1	58.4	250	4280	20.1	15.1	-	16.2	0.39	
		2	57.5	250	4350	21.6	16.2	16.2			
		3	57.3	250	4360	22.1	16.6	16.6			
		4	61.4	250	4070	15.3	11.5	-			
		5	60.2	260	4320	21.0	15.8	15.8			
		6	57.7	260	4500	26.4	19.8	-			
		7	58.0	260	4480	24.1	18.1	-			
		8	114.6	500	4360	21.9	16.4	16.4			
		9	113.6	500	4400	29.0	21.8	-			
		10	114.4	500	4370	22.0	16.5	16.5			
		11	57.7	250	4330	20.8	15.6	15.6			
		12	57.3	250	4360	22.0	16.5	16.5			
		13	53.0	250	4720	33.5	25.1	-			

Primary data placed in column 4 of table, signifies signal passing through the cross-section of the column (t, mks). Column 5 contains the measurement data base of sonic test (B, mm). Column 6 contains the results of determining the speed (V) completion of ultrasound on the basis of the (B):

153

$$V\,(\text{m/sec.}) = B(\text{mm}) \,/\, \text{sec.(mks)}. \tag{10.18}$$

In the found value of the velocity V, using empirical dependence [10.1] one can get:

$$R(\text{MPa}) = 0,802 \cdot e^{0,00129V(\text{m/s})}. \tag{10.19}$$

Here R is the cube strength of the concrete columns. The value of calculations by formula (10.18) is placed in column 7 of the table. The transition to the characteristic values of the strength R^H of concrete is carried out by multiplying the values of R in column 7 by a factor $\varphi_b = 0.75$ [10.10]. The results are placed in column 8 of the table. You can see that the range of values of R in column 7 has maximum (R^Hmax $= 25.1$MPa) and minimum (R^Hmin$= 11.5$ MPa), and differ significantly from those of the main array of the values. But we still do not know, whether they are the consequence of a gross error, or they reflect the actual dispersion properties of concrete study of the column. Thus, it is necessary to solve the question of the legality culling of these values.

A) Let us consider the possibility of rejection $R^H = 11.5$MPa (see table. 1) when value from the analysis is excluded, putting the number of $R^H{}_n$ equal 12. Then the mathematical expectation will be

$$m(R^H) = 17.8 \; MPa. \tag{10.20}$$

Since the value of the standard deviation is unknown beforehand, it is estimated approximately in the results of the column 8 of the table, i.e. instead this is used instead of her empirical standard:

$$s(R^H) = \left(\frac{1}{(n-1)} \sum_{i=1}^{n} \left(R_i^H - mR^H \right) \right)^2 = 3,00 \; MPa. \tag{10.21}$$

Compare the absolute value of the difference $|R^{H*} - m(R^H)|$ with the value of the "s":

$$t = \left| \frac{R^{H*} - m(R^H)}{s} \right| = \left| \frac{11,5 - 17,8}{3,0} \right| = 2,7. \tag{10.22}$$

We can compare the value of t with a critical value $t_n(P)$ (table 10.2, [10.2]).

B) Consideration of the possibility of rejection $R^H = 25.1 \; MPa$ (see table 10.1) in accordance with presented in A) methodology (for $n = 11$) gives a value of $P=0.994$, that is, the value of $R^H = 25.1 \; MPa$ is also rejected.

154

Table 10.2

The critical values $t_n(P)$ relations (10.22) for the rejection of the «jump-out» values of t.
(n is the number of acceptable results, the P – reliability of the output)

n	P				n	P			
	0,95	0,98	0,99	0,999		0,95	0,98	0,99	0,999
5	3,04	4,11	5,04	9,43	20	2,145	2,602	2,932	3,979
6	2,78	3,64	4,36	7,41	25	2,105	2,541	2,852	3,819
7	2,62	3,36	3,96	6,37	30	2,079	2,503	2,802	3,719
8	2,51	3,18	3,71	5,73	35	2,061	2,477	2,768	3,652
9	2,43	3,05	3,54	5,31	40	2,048	2,456	2,742	3,602
10	2,37	2,96	3,41	5,01	45	2,038	2,441	2,722	3,56
11	2,33	2,89	3,31	4,79	50	2,030	2,429	2,707	3,532
12	2,29	2,83	3,23	4,62	60	2,018	2,411	2,683	3,492
13	2,26	2,78	3,17	4,48	70	2,009	2,399	2,667	3,462
14	2,24	2,74	3,12	4,37	80	2,003	2,389	2,655	3,439
15	2,22	2,71	3,08	4,28	90	1,998	2,382	2,646	3,423
16	2,20	2,68	3,04	4,20	100	1,994	2,377	2,639	3,409
17	2,18	2,66	3,01	4,13	∞	1,960	2,326	2,576	3,291
18	2,17	2,64	2,98	4,07					

Notes:

1. Linear interpolation on argument n may give an error of up to 10^{-2} at $20 < n < 60$ and error of up to 10^{-3} at $60 < n < 100$.

2. For $n > 100$ critical values $t_n(P)$ with an accuracy of up to 10^{-3} can be calculated by the formula

$$t_n(P) = t_\infty(P) + \frac{t_{100}(P) - t_\infty(P)}{n} 100 . \qquad (10.23)$$

For $n = 12$ (table 10.2) $2.29 < t = 2.7 < 2.83$. Linear interpolation gives the value of $P = 0.97(97\%)$. Thus, with the reliability, greater than $P = 0.97$ we can say that «jump-out» the value of $R^{H*} = 11.5 MPa$ contains a mistake and it should be excluded from further consideration.

C) In Further consistent consideration of the values of $R^H = \{21.8; 19.8; 18.1; 15.1\}$ MPa (see table10. 1) in accordance with presented in A) methodology leads to a conclusion about their misguided, and they also rejected.

D) Consideration of the possibility of rejection $R^H = 15.6\ MPa$ (see table 10.1) in accordance with presented in A) methodology (for n = 6) gives the value $t = 2.33$ and $P=(n=6;\ t=2.33) < P(n = 6;\ t = 2.78) = 0.95$, that is, we do not have sufficient confidence in the fallacy of the values of $R^H = 15.6\ MPa$. It remains in the package (column 9 of the table 10.1).

155

According to column 9 of the table 10.1 and using formulas (10.20) and (10.21) the following parameters are determined

$$M(R^H) = \frac{1}{n}\sum_{i=1}^{n} R_i^H = 16,2 \ MPa;$$

$$S(R^H) = \sqrt{\frac{1}{n-1}\sum_{i=1}^{n}\left(R_i^H - m(R^H)\right)^2} = 0,39 \ MПa.$$

The design resistance of concrete to axial compression

$$R_b = M(R^H)(1 - t_\alpha \cdot s(R^H)) \qquad (10.24)$$

Here t_α – parameter Student's t distribution (table 10.3).

When the number of tests is n = 7 then class of concrete on compression (by formula (10.7)): $B = M(R^H)(1 - t_\alpha \cdot s(R^H)) = 15.5$. In accordance with the classification of [10.11] concrete is taken by type B15. The design resistance for axial compression is taken by [10.11, table. 13].

Table 10.3

The values of the Student's factor t_α at $P = 0.95$ (single-sided limit)

Number of tests	t_α	Number of tests	t_α
1	6,31	11	1,80
2	2,92	12	1,78
3	2,35	13	1,77
4	2,13	14	1,76
5	2,01	15	1,75
6	1,94	20	1,73
7	1,89	25	1,71
8	1,86	30	1,70
9	1,83	40	1,68
10	1,81	∞	1,64

And, finally, in the procedure for discussion: about the initial reserve of structural strength. The discussion will hold in a deterministic setting.

The structure should be designed to provide for a high level of it safety in terms of strength during operation. Assessment of the safety of the structure under the terms of strength must show that within the (scheduled) the period of service in the anticipated operating conditions emergency situations are almost impossible.

Where does the initial reserve of structural strength come from? But then in the optimal design should, on the one hand, maintain the numerical equality between influencing the design factors and the ability of structures to resist this influence. Thus it turns out that the initial reserve of strength of load-bearing structures does not exist, as the design also accepts the computational load, which wasn't implemented at the same time. On the other hand, the initial strength reserve should exist. Explanation of the presence of initial reserve of structural strength and introducing to the design a number of factors (model factor and other partial factors), is not correct in our opinion. Partial factors take into account the statistical nature of the various values at a given level of reliability of the design results.

The challenge is to ensure the characteristic duration of the operational life cycle (without capital repair) of the building t_H for a given function of changes the ability of the building to perceive loads in time (10.8). It is necessary to find the initial strength reserve already at the stage of design, i.e., to determine the necessary size of the initial strength factor k_0. And, already thinking in the design of this task, in the process of exploitation of the building (structure) it is possible to evaluate the residual life of the building on the results of conducted surveys, and in this case to define the remaining time of its safe operation.

$$t_{RS} = t_{\text{н}} - t = t_{\text{н}} \left[1 - (k_0^2 - k^2)/(k_0^2 - 1) \right]. \qquad (10.25)$$

The value of t_H is the codified value and is determined by the corresponding standards documents, for example [10.12–10.14]:

It should be noted, however, that this kind of classifications in most of them was created by institutional organizations on different methodological basis, which do not agree with on terminology and are not united by common methodology.

The value of k_0 depends on the type of buildings and on the specific values of «k» at the appropriate time (as a rule, in moments of successive surveys). In determining the optimal value of the parameter k_0 two approaches can be used:

– identification of another boundary condition linking k_0 with variable parameters. Perhaps, it exists, but it hasn't been successfully formulated yet;

– determination of the values of $k_i = f(t_i)$ on the results of previously conducted surveys ($i = 1,2,n$), and then obtain the values of k_0.

Of course, the second approach at the initial stage can give significant errors in determining the value of k_0, but all the same it is the reference point, which is gradually with the accumulation of evidence will be in the limit closer to the deterministic value. At the present time, the lack of attempts to standardize k_0 leads to a very large overvaluation of the initial strength reserve of building structures. To illustrate data, published Duzinkevich M.S., Lysov D.A., Chayner H.E. in [10.14] can be used. From this data it's clear that the initial reserve of the strength of the interior walls of the five-storey residential houses of a series of l-510, l-515 was 4.00-4.35.

Here is an example of the k_0 and t_{RS} calculation for the building. For example – the building belongs to the 1st group of solidity and has been operating for 50 years. According to the results of previously conducted surveys the average value is $k_0 = 1.32$. In the regular examination of the building at the moment of time $t_1 = 50$ years is set $k_1 = 1.23$. According to the formula (10.25) $t_{RS} = 150 [1 - (1.32^2 - 1.23^2)/(1.32^2 - 1)] = 103$ years. Thus, under the terms of the preservation in time of the functions of the building perceive the design load; the operation of the building can be regarded as satisfactory. If we change some of the provided examples of the value of k_1 and let the results of the survey be ($t_1 = 50$ years) k_1 will equal 1.15. In this case the residual life time $t_{RS} = 65$ years. That is, the remaining lifetime of buildings decreased and we can make the conclusion about the low level of operational services.

The above approach to the definition of the residual (on time) of the resource of the building is more of a methodological character on the whole. It is expedient to evaluate the residual life and for it to draw conclusions for each of the bearing structures of the building separately. Determination of the necessary volume of repair work can be performed using the formula which allows getting integral assessment of the state due to load-bearing structures of the building according to the actual situation or according to the residual resource.

$$k = \sum_{n=1}^{i} \alpha_i \cdot k_i, \qquad (10.26)$$

where k is the integral assessment of the resilience of the entire building;

k_i – reserve assessment of the strength of the i-th bearing structure of the building;

i_f – weight factor of the i-th bearing structure of the building.

Values α_i for individual load-bearing elements can be taken according to the standard documents depending on the type and solidity of the building.

The approximate unit weight of load-bearing elements α_i for large-panel 5-storey residential house is given in table 10.4.

Table 10.4
Design unit weight (in cost terms) elements of large-panel the 5-storey apartment house

Elements of building	Unit weight %
Basements	6
Walls	55
Partitions	9
Floors	16
Roof	8
Roof covering	3
Stairs	2
Balconies	1

Ch.11. Problems of Codified Reliability Methods, Rules of Operation, Survey and Prediction of Service Life for Structures

Kozachek V.G.[*]

Sec.11.1. Historical Notes

In a number of international organizations in the programs for development and improvement of a unified codified base in the field of building construction, they are working going on introduction of probabilistic approaches to the codes of building designs [11.1-11.13]. The fundamental documents in the field of reliability of building structures are being developed, as well as practical guidance on the survey and assessment of the reliability of particular types of structures (bridges, etc.). Data base of the variability of the parameters of structures and loads is being accumulated and systematized which is necessary for probabilistic designs. For the sake of justice it should be noted that many national monographs and developments in the field of theory of reliability at some point of time were ahead of foreign publications.

There are the works of V.V.Bolotin, N.S. Streletskyi, B.I.Snarskis, A.R. Rzanitsyn, and later – V.D.Raizer, A.P. Kudzis, V.M. Bondarenko, V.P. Chirkov, etc. However, as it often happens the works mentioned above were not used for practical application, while this direction has been constantly evolving abroad.

In the field of fully probabilistic methods of design the problem is not mostly in the theoretical part but in the practical implementation, although solutions for the most complicated tasks on the basis of the probabilistic approach (for example, for statically indeterminate systems with the joint account of the physical and geometric nonlinearity, long-term processes do not have the properties of additivity and etc.) have not been developed yet. The reasonable use of fully probabilistic approaches to real design or for the practical evaluation of the reliability of existing structures is a matter of future. Therefore, there still needs to be performed urgent improvement of traditional design methods with the use of reliability factors on the basis of their refinement with the involvement of the theory of reliability. However, it is clear that the establishment and differentiation of specific numerical values of parameters of reliability in the codes creates on one hand the scientific basis for the design of structures of varying reliability, but on the other - significantly complicates the procedure of selecting reasonable values of the factors that ensure the

[*] Ataev Institute NIIPTIS, Minsk

required reliability and reveals her vulnerability because of the conventions of some of the theoretical provisions.

With the introduction in the 1955 codes regulations, the method of limit states, allowed a more differentiated take into account the variability of loads and resistance of structures. To a certain extent, also align the reliability of some elements of the building, the statistical data for the majority of initial parameters formulas was missing, and reliability theory was not ready for decision of practical problems, especially for composite structures, such as reinforced concrete. Therefore, the safety parameters in the design formulas asked almost willed manner, so that the final results of the design is not very different from the other methods. It was assumed that when the statistics will be saved, it will be possible from probabilistic positions justify the numerical values of the safety parameters. As we can see, over the past 56 years, the real results were obtained only recently, but they apparently cannot be unequivocally described as quite successful.

Since 1955, the general view of the inequality that characterizes the conditions of provision of bearing ability of designs remained unchanged:

$$E_d \leq R_d \qquad (11.1)$$

where E_d – calculated (taking into account the possible variability in the big side) values of forces;

R_d – calculated (taking into account the possible variability in the smaller side) limit resistance design.

Not considering the left part of the inequality (11.1) it should be noted that the form of the records of expression for R_d changed several times (sometimes conceptually). So, in the development of Russian Code (SNiP P-B.1-62) condition (11.1) was originally written in the form:

$$E_d \leq mR \{S; R_{kc}; k_c; m_c; R_{sk}; k_s; m_s\}, \qquad (11.2)$$

where R_k (R_{sk}) – characteristic values of resistance of concrete and reinforcement (taken equal to the average values);

$m; m_c; m_s$ – model factors of structure, concrete and reinforcement;

$k_c; k_s$ – material factors of concrete and reinforcement.

The numerical values of the material property factors of concrete and reinforcement have been established for the transition from the average strength directly to the current resistance with exceedance probability of 0,999, i.e. (under the normal distribution) on the basis of the rules of

$3s$ ($k=1-3\dfrac{\sigma}{R_k}$).

It was assumed that the model factor takes into account the causes, influencing the work of the whole structure, for example the spatial work, which for any reason cannot be taken into account directly in determining forces; inaccuracies (error) the design diagrams, formulas etc. The peculiarities of manufacture and operation of structures can also be taken into account with the help of this factor, for example a guaranteed level of supervision, etc. The model factors (the working conditions of material- m_c, m_s) take into account the peculiarities of overloading of the structure in relation to a particular design, as well as to the functions, which the property of material in this design performs in certain operating conditions. In the final version of Russian Code (SNiP P-B.1-62) factors k, $m_{c(s)}$ are not explicitly included, but were taken into account in the tabular values of the calculated resistance. Factor m was not included into design dependences of the code.

In developing the subsequent codes (1975 and 1983.) this approach was adjusted. In particular, it is proposed to divide factors into two groups: those, which at this stage can be estimated as statistically significant (strength characteristics of the material); and those which are of a statistical nature, but has not yet amenable to probabilistic assessment (differences in the strength of concrete in structures on the object and the strength of the control samples, the variability of the size of cross-sections, etc.). The latter factors invited to take into account the factor of safety (reliability) $\gamma_{c(s)} > 1$. The numerical values of these factors is installed so that the values of the design resistance of concrete and reinforcement, some with a view of statistical and non-statistical factors $(R = \dfrac{\overline{R}(1 - 1.64C_v)m_i}{\gamma_c}$), is not very different from designated in the previous codes directly through the middle of the strength of $(R = \overline{R}$ -3s) in the «average» rates of variation. It was considered that in such a two-stage transition from average to the designed characteristics of the material ($\overline{R} \rightarrow R_H \rightarrow R$), even with the small s (c_v), there is a reliable gap between the estimated and average (expected) strength of the material for the creation of the required total reserve of the whole structure. All other features should take into account capacity of materials property, as previously, with the help of numerous (12!) model factors m_i.

Thus, the basic design characteristic of the materials became characteristic resistance, guaranteed by the manufacturer with exceedance probability of 0.95. However, the reliability of the structures were given and estimated very roughly, by analogy with the safety factor in the method of destroying loads. However, already then it was clear that such an approach required further improvement, as it happened in the exa,ple when a conditional reliability factor changed in wide limits - from 1.25 to lightly reinforced concrete bending structures, to 1.6 close to the central compression. It is clear that the minimum of these values was insufficient for the creation of «enough reserve» and in the codes had to use some of the parameters which are different for comparison with the experience and for the design. These parameters were adjusting the design formulas, but were wrongfully named "model factors" (for example, $m_{a4}(\gamma_{s6})$). As the development of the design theory the calculation formulas changed and the need for the application of such factors disappeared (for example, with the use of real diagrams of deformation of materials).

In modern international documents that regulate the basic rules of the design of structures, several forms for recording expressions for the R_d are being offered, which differ among themselves not only in writing but also in the sense of [11.1-11.4, 11.10 and 11.11]. For example, in [11.4] written
   – the general form

$$R_d = \frac{1}{\gamma_{Rd}} R\{\eta_i \frac{X_{Ri}}{\gamma_{mi}}; a_d\} \tag{11.3}$$

   – the simplified form

$$R_d = R\{\eta_i \frac{X_{ki}}{\gamma_{Mi}}; a_d\} \tag{11.4}$$

where R_d is the design value of resistance of the structure (cross-section);

   X_i – the characteristic value of the properties of the material;

   a_d – design value of the geometric parameter;

   γ_{Rd} – factor, taking into account the errors of design formulas and the spread of geometrical parameters, if he is not considered by a more accurate way;

   γ_{mi} – factor of safety «by material»; the same thing and γ_m, but also includes accounting errors of the estimated model and variability of geometrical parameters, $\gamma_{M,i} = \gamma_{Rd}\gamma_{mi}$; η – correction factor.

The simplified form of writing, proposed in [11.1], is presented in two variants (11.5) and (11.6), which use nominal values of geometrical parameters, and the η was already included in the γ_M

$$R_d = R\{\frac{X_{ki}}{\gamma_{Mi}}; a_{nom}\} \tag{11.5}$$

$$R_d = \frac{1}{\gamma_{Rd}} R\{X_{ki}; a_{nom}\} \tag{11.6}$$

In the regulations (building codes) of the United States, Canada and some other countries the same formula (6) is used in a simple form [11.7]. It is established, that the factor γ_{Rd} should be adjusted (calibrated) in conjunction with the load factor so that the rules of probability of indestructibility of the design were provided with the minimum reserve. Strictly speaking, each safety factor takes into account the variability of only one input parameter and, from a mathematical point of view, because the reliability of the design is a function of many variables, each factor is determined by the partial derivative of the function at the appropriate argument. In the codes of the majority of countries the form (11.5) is used. Although it is obvious that the adoption as a basis, due to the design or to the assessing the suitability of the operation of existing structures, the semi probabilistic approach but the form (11.6) is the more acceptable as so-called design characteristics of materials are not present, as such.

Sec. 11.2. Recommended Methods of Calibration of Safety Factors

Let us consider the recommended methods of calibration of parameters of safety. It is recommended in [11.1] to calibrate the safety factors so that the design index of the reliability β is very close to the codified one βt. It is obvious, that in the various design situations optimal values of γ_{mi} will be different. For the valid of destination of unified values of the partial factors numerous studies should be conducted enforce the types and parameters of structures, types and schemes upload, etc. [11.1, 11.8]. The calibration procedure should allow, in general, to solve the optimization problem, which gives the «best» combination of all of the reliability factors as on the loading side, and on the side of a structure with account of codified (close to the real) statistical characteristics of all variables. In fact, this is the most complicated optimization task, which you can try to solve by different methods, for example, using the methods of FORM or SORM [11.1, 11.3, 11.9, 11.11, 11.12], as well as with the use of mathematical programming methods. A complete mathematical experiment should be taken with the use of the proposed with codes linear and nonlinear models of structural resistance. Such work is conducted in the framework of the European bodies, but it is still in the initial stage, because it considers only the simplest kinds of stress states, loading, etc. (see for example [11.10]). In practice initially parameters of cross-sections are being calibrated approximately, and, fixing those, their factors of reliability and load combinations are being calibrated for obtaining a general index of reliability, close to the codified.

While it is early to make any generalizations, it is clear, however, that the more in the factors are in the formulas, which need to calibrate, the more uncertain solutions will be obtained. It is known that in optimization problems close to the optimal value of the objective function (at rather gentle curves) you can get many sets of combinations of the target factors, giving close to the «best» result. In is not quite clear the very ideology of improvement of the process of clarification of reliability calculation by gradual accumulation of information about the statistical variability of the growing number of variables and, accordingly, taking them into account in designing dependencies. Strictly speaking, every such stage clarification must be accompanied by a full revision of the whole set of partial factors, as they all shared influence on the final design reliability. In the body of the approximate formulas with numerous empirical dependencies, designed characteristics are intertwined in a tight ball, repeatedly re-recorded .to eliminate this «double taxation» we have to enter different values of some parameters for comparison with testing and for further design implementation

The problem of a real reliability of a structure is confusing enough and pseudo scientific. In this case it is harmful, because it only creates the appearance of «reliability» while on the surface there are some obvious, but unresolved issues. For example, it is illogical to believe that exceedance probability of design resistance of materials in the formulas of the type (11.5) is identical with the exceedance probability of design resistance of cross-section as a whole and especially of the whole structure, made of these materials. No doubt that the probability of the coincidence in one cross-section of the minimum characteristics of concrete and reinforcement is much smaller than the on in separate materials. On the left, in the loading side of the expression (11.1) a similar circumstance is taken into account by factor (ψ). Cross-section of reinforced concrete element is internally «statically indeterminate» and the assessment of its reliability must be in the general case done with the use of the approaches being developed for multi-element systems. The solution to problems with the parallel connection of elements is preferred, apparently, (redundancy system), allowing to take into account the increased reliability of such systems, because the failure of one of the elements only leads to a redistribution of efforts without loss of efficiency. The reliability of the system is substantially higher the reliability of any individual element. Obviously, the lack of theoretical solutions of this problem for reinforced concrete cross-sections should not be a reason for ignoring the properties of their increased «survivability».

The property of the cross-sections of several materials can be logically taken into account in the value of the coefficient (γ_{Rd}) similar in form, if the expression (11.6) for the right-hand side of is being used (11.1) (but a few of the other content in comparison with the factor m in equation (11.2)). This factor can be expressed in a multiplicative form for ac-

165

counting purposes, including other factors, providing generally required reserve on the part of the design. There is a possibility in logical way to adjust the excess reserves from the existing structures if the value of γ_{Rd} in expression (11.6) is qualitatively calibated. It allows to eliminate contradictions, noted in [11.13] between the results of calculation of structures on the codes supportive of it's overloading and its actual good condition. Thus, the results of such calculations are close to the results obtained by fully probabilistic methods (for example, the methods of statistical modeling); especially if they take into account the statistical variability of a larger number of variables than is generally accepted.

Obviously, when calibrating the only one reliability factor for cross-section of structure γ_{Rd} algorithm of calibration will be simplified and the dispersion of the results will be less than with the traditional approach. It is necessary also to take into account the fact that, under other equal conditions, the reliability of the structure, for example, beams with variable moments diagram (for example, triangular), will be significantly higher than with the permanent ones, because again the same probability of coincidence of minimum materials property and the maximum value of the force in the first case, less than in the second. The expression (11.6) can be used for the design of the nonlinear deformable structures, where the load carrying capacity (the maximum load on the strength or stability) is determined using the estimated models of codified characteristics of materials, and then finally entered the total reserve factor is being entered. Large design experience with this approach has been gained in the former USSR in the design of standardized frameworks of industrial buildings [11.14].

Thus, to regulate the reliability of structures as a whole it is also more convenient and logical to do so by calibration of a single factor γ_{Rd}, and to not make a modification of the reliability material factors, which, as already noted, for reinforced concrete structures has no clear physical meaning. In connection with the mentioned above it is not quite correct to regulat Annex A to the code [11.2] presented in detail. Proposal consists of reduction of the γ_{mi} for concrete and reinforcement depending on the level of control over the production of reinforced concrete structures and the more depending on certain specific variability of the geometric parameters of the cross sections. High stability of the nature of the materials is usually defined by the fact that in order to receive guaranteed strength of a material in the production, the average strength is being reduced. The integral accounting of the level of control over the production or over the exploitation in the design or survey with correction of γ_{Rd} factor is also

166

more logical than the correction the reliability factors of concrete and reinforcement. It should be considered that those foreign proposals for specific accounting variability of the source parameters in formulas of the design are based on the statistical data obtained in the conditions of high culture of manufacture and are often unacceptable in our conditions. On the other hand, the situation seems to be false as we have the ideology of ensuring geometric accuracy in the construction industry. The values of the technological tolerances of the manufacture and assembling of the structures in national standards are established (in several times less than abroad, in complete isolation from the real opportunities of the specific production and, respectively, and practically are not controlled and are not regulated. Nominal dimensions are assigned to model series, or to individual projects on the basis of the strength calculations, principles of unification, architectural considerations, etc., but not confirmed by the calculation accuracy of the assembly of the building, which, by the way, is regulated by the functional documents. As a result, everywhere the functional tolerances are not being observed when it comes to geometric accuracy of structural assembly, as an example of the parameter of depth exchanges of supporting plate coating on the roof trusses. These joints have to be constantly enforce . In the system of ISO standards nominal dimensions of structures are the resulting parameter, appointed by the project on the basis of the design of the geometric accuracy, performed with account of real, guaranteed by the manufacturer, tolerances for manufacturing and installation.

Thus, it is proposed the use of a common entry for the determination of the resistance of cross-section (structure) in codes, both for the design, and for the assessment of existing structures in the form (11.6), but in a slightly modified form:

$$R_d = \frac{1}{\gamma_{Rd}} R \{\eta_i X_{ki} ; a_{nom}\}, \qquad (11.7)$$

where in brackets along with the nominal of geometrical parameters (a_{nom}) there are engaged guaranteed, statistically controlled by the codes the resistance of materials, properties X_{ki} together with model factors η_i, which take account of the specific characteristics of materials that do not respond at this stage of development due to the design theory of analytical description. Reliability factor γ_{Rd} may consider in multiplicative form the variability of the geometric characteristics (γ_h), the inaccuracy of the estimated model (γ_{mod}), the combination factor of materials (ψ_R); the factor (γ_ℓ) which can in an integrated approach consider the guaranteed level of control over the manufacture of products (but not material!) or consider the exploitation, as well as the correction factor(γ_c), providing the necessary overall reliability of the structure (finally calibrated part of the γ_{Rd}):

$$\gamma_{Rd} = \gamma_h\, \gamma_{mod}\, \gamma_\ell\, \psi_R\, \gamma_c. \qquad (11.8)$$

The meaning of factor ψ_R should be variable and be set $\psi = 1$ in the case of the central compression or tension, and for other stressful conditions (with stress gradient, etc.), where the interplay of concrete and reinforcement in ensuring the overall reliability of the design becomes fairly complex form, the value of ψ_R should be less than 1. Sound calibration of the factor γ_c will bring together the results of semi probabilistic and fully probabilistic design method, removing the existing contradictions, arising in the diagnosis of structures and avoid quite far-fetched (in the light of the above considerations) modification of partial reliability factors on materials in the process of examination of the proposed in work [11.5].

In the design of existing structures the consideration of the actual variability of the properties is an abstracttion, but is an actual design, and on its basis the modification of the parameters of safety in the practice of the survey is significantly complicated by purely technical difficulties. Specialists of the survey are well aware that it is practically impossible to perform multiple (sufficient to obtain not too much of a bias in the statistical processing) measure many parameters. This applies, for example, to the height of the cross-section of floor slabs and coverings, and girders with bearing up the plates on the console (series II-60; II-04; 1.020 etc.) to a degree of reinforcement corrosion and destruction of concrete in depth section, which in various situations is significantly different in length (height) of the element, and on different planes design depending on their "proximity" to the aggressive impact, etc. It is practically impossible to control the quality of the bathroom welding in the monolith joints and, accordingly, take into account the real flexibility of the joints of structures, etc.

There are a lot of not solved outstanding methodological issues in determining the strength characteristics of concrete. For example, cutting a sufficient number of kerns from the designs cover is not technically feasible. By non-destructive methods of control (also known conventions, connected with the necessity of their connection with destructive testing, etc.) the coefficient of variation in the regulations of the Russian Federation should define the vector sum of the coefficients of variation of instrument readings and variations of this method [11.15]. The resulting value of C_v turns out to be quite large, and the calculated resistance may be lower than the rated. Often determination of the strength of concrete in the compressed zone (which may vary significantly from the strength in the extended zone) is technically impossible, because it is hidden by coupled structures. The problem is, that we get in a survey the actual data on a specific date, and the objective is very difficult enough to forecast change of strength, weight, the deformation and other characteristics of the concrete, the «pie» of enclosing structures, etc. For example, the weight and thermal properties of insulation depends on the humidity, which can significantly change due to natural wear and untimely repair roll coating and its uncontrolled mechanical damages, which is common in our conditions. If there is objective data on the variability of a larger number of parameters of simple structures with simple loads it is possible in principle to assess reliability on a probabilistic basis, which, however, is valid only for the specific date of the survey. Methods of reasonable forecast of the change of reliability «for the future», suitable for practical application, have not been developed.

This points out that there is a need for a sufficiently large reserve of bearing ability of structures for the "unforeseen circumstances", which in the framework of the semi probabilistic method of design, using the case of (11.7), can be regulated in the small limits by the safety factor of cross-section γ_{Rd} (such as guaranteed by the owner of the level of the subsequent operation), but not modifications of the factors of the materials safety. In particular, in the USA Building Code Regulations [7], some adjustment of the general "safety factor" is allowed only in the case of the simplest of stress states and at presence of the full unbiased information of the geometrical parameters of a cross-sections and strength characteristics of materials. The possibility of some reduction of safety factors for the individual loads, when designing the structures of existing buildings with a view of the expected residual term of their service, should be explored

The opinion of the authors seems to be reasonable in[11.6], which considers that in spite of the fact that some experts are inclined to use different values of the safety factors in the design and survey of reinforced concrete structures, there is no single opinion and acceptable methodology at the present time on this issue. (Engineering experience and qualification of the specialists of the expert organization in such a situation, is the most important condition for making the right decisions when assessing the suitability of existing designs. It is difficult not to agree with this statement.

In connection with this, we note some of the existing differences in the understanding and definition of certain terms. In particular, one should not equate the concept "the technical state" and "the carrying capacity". Technical state, as "the totality of features, which characterize the degree of compliance of the building to draft or codes" is a more general term, as it applies to all, and not only to the load-bearing elements of the building [11.20, 11.21, 11.23-11.25]. Technical condition of load-bearing structures has not always having been linked to existing or anticipated loads. For example, at the stage of general survey, when assessing the condition of structures by external characteristics, status at the time of the survey can be characterized as satisfactory, although, the owner is planning to increase the load and the carrying capacity of the same design may not be ensured. Thus, the technical condition characterizes the compliance of the design requirements (including stress) at the time of the survey. The final evaluation of the conformity of existing structures (or anticipated) in operating conditions, it's the forecast that must be carried out only after clarification of all initial data with a detailed survey and design [11.28].

(The above approach is provided by general schemes of estimation due to technical condition of structures, developed by us earlier [11.20] and in informal form included in [11.23, 11.28]. It is obvious, that for the implementation of the provisional classification of defects and technical conditions of the external signs in accordance, for example, with [11.23] the experts should have sufficiently profound knowledge and experience. There are often T situations when you can confine to the general survey. For example – when it's technically impossible to conduct a detailed study of operated buildingsor when sufficient funds from the owners are absent, or when the serious defectsa are absent. Moreover, it is possible on its basis to take a decision on the compliance of the structures to requirements, including the assessment of their bearing ability and operational suitability, subject to a number of obligatory conditions and without performing designs [11.28]. Such situations are provided in the European standards [11.3].

Nevertheless, the experience of operation of structures shows that any building must be, at least, once examined in detail, and later - the necessity of the survey, the specific timing and level of detail should be clarified on the basis of the results of surveillance of the building in the process of exploitation. The same can be said about the necessary maintenance activities in the conditions of a lack of funds for all necessary "preventive" measures, which were established earlier in the former

USSR the system of preventive maintenance. Such a system, by the way, at the present time is successfully functioning in many countries. Thus, the regulations of [11.23-11.28] imply that the proposed scheme for surveillance of buildings with exploitative services which periodically carried out the general surveys (dates, which must be, as a rule, reconciled with the terms of changes in the passport of the building) and will provide effective monitoring and timely implementation of maintenance activities for extending the useful life of buildings.

Sec.11.3. Service Life and Durability of Buildings

The term "working life" is the key to the design, assessment and prediction of durability of buildings and structures. It should be noted that the "working or service life" in the national and foreign design regulations was not codified in recent years. It was believed that the design performed according to the both groups of limit states and compliance with regulatory requirements (in terms of parameters of the protective layer, anti-corrosion coatings, etc.) provides «sufficient» durability. The exception, apparently, was the code (SNiP P-22-81) "Stone and reinforced masonry structures", where the requirements to the materials and mortars are aligned with the expected lifespan of buildings (25, 50 and 100 years). In this connection it should be noted that in the national regulatory and guidance documents are not always lacked the regulation of terms of service.

In the former USSR in the field of construction, due to the work of Soviet scientists [11.16, 11.7, etc.], for the first time in the world on the basis of the group's standards of "Reliability of the equipment" in the early 60sof the last century a systematic approach has been developed to ensure the reliability and durability of buildings and structures. Durability is characterized by the time during which the facilities (with breaks for repair) of performance will not be reduced below the level specified in the draft or in the regulations. It is determined by the full term of service for permanent-set with the capital repair of structures such as foundations, walls, columns, reinforced concrete floors, etc. As a rule, the date of occurrence of the limit status, criteria of which can be defined not only by carrying properties of structures, but also economic, aesthetic and other considerations. Depreciation of structures and buildings in general is the loss of the original operational qualities in time. This process is inevitable, and the main objective in the operation consists in timely repair, replacement of structures with small terms of service, strengthening of the main structures, etc. Even in ideal conditions, the rate of operational quality levels decline with time due to the natural ageing of materials, and their interaction with the environment, etc. (Fig.11.1).

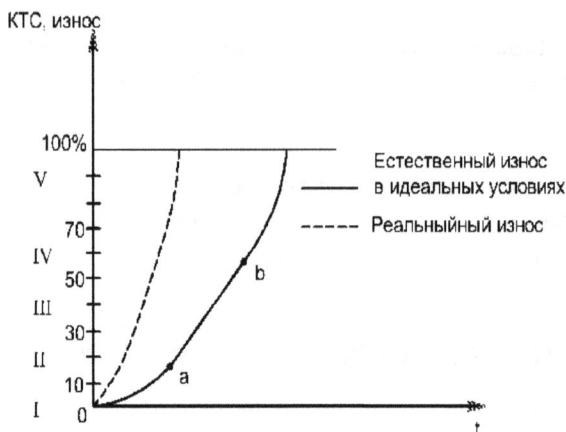

Fig.11.1. Scheme of deterioration (of a technical condition of structures by the time). Category of a technical condition (CTC): *I* – operable; *II* – working; *III* – limited working; *IV* – an unusable; *V* – pre-collision (limit)

In real conditions, particularly when there is a low initial quality of materials, works, and violation of the rules of operation, etc. the intensity of wear increases significantly, and durability of structures can be reduced in several times if there is no strict observance of the established volume and terms of maintenance and repair work. Depreciation with the passage of time after the warm-up period (0-a) for some time was stabilized (a-b), and then gradually speeded up considerably after reaching the age structures of 80% and more from the full term of their service (see Fig.11.1). The accident rate of the buildings directly connected with their quality. Despite the large number of domestic publications on statistics of accidents, their objectivity often raises the doubts, and not only because of the lack of a clear system of accounting and analysis of failures and accidents. Unfortunately, the investigation of the causes of accidents is sometimes subjective, as are the departmental commissions and therefore contained findings in the official documents are often of a subjective nature. Thus, the statistics of accidents usually is being usually overstated in the direction of "design errors",because in many cases the particular manufacturing defect is virtually impossible to detect in the ruined structure without deep research, and there is usually no time for that,, but in any design or method of calculation it is easy to identify those or other defects. Interesting data on accidents of steel structures (594 cases) in the post-war period in Germany is shown in the independent study [11.34] (see table 11.1 and 11.2). It's obvious here in table 11.1 that the the number of accidents happened not in special facilities, but in mass –

produced buildings which are usually characterized by great variety of design solutions, operating conditions and quality of implementation.

Table 11.1

The distribution of accidents by type of buildings and structures

Type of structures	Number of accidents	Interests
Buildings	223	39.5
Road bridges	79	14.0
Railway bridges	80	14.2
Engineering structures	40	7.1
Crane structures	88	15.6
Other	22	3.9

Table 11.2

The distribution of emergency cases according to the period of operation

The duration of the exploitation to collapse (years)	Number of accidents	%
1-10	142	32.4
11-20	87	19.9
21-30	38	8.7
31-40	17	3.9
41-50	33	7.5
61-70	21	4.8
71-80	29	6.6
More then 80	9	2.5
Not established	51	11.5
The whole	430	100.0

The main number of accidents occurs in the first 10-15 years of operation, during which gross errors of design will appear and production activities (table 11.2). Later in life, until the age of ≈ 70 years, there is a period of relatively uniform distribution of the resource refusal of structures. Significant reduction of the accident rate of the "old" buildings in Germany is explained on one hand by their relatively small number of due war destruction, on the other by a large \reserve of durability of old buildings, a more thorough supervision over them and timely disposal or qualitative restoration of historical buildings.

You can select «optimal» durability- service life of buildings (mass construction), which it is still appropriate to (from economic point of you) to restore. In excess of that period, especially if the necessary timely repair (was default) what does it mean?, the restoration costs are increasing dramatically and may exceed the cost of assembling a new building (Fig.11.2).

The corresponding methods of calculation of the economic expediency of restoration of buildings are used by the specialized institutions in the design of capital repairs, but they are far from perfect.

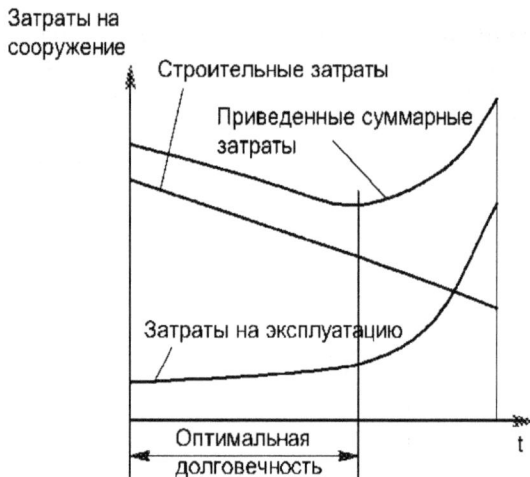

Fig.11.2. The scheme of change of costs for operation in time

The simultaneous consideration of the physical and moral ageing, which has its own specifics, is important especially in the commercial buildings, because in the modern conditions of frequent modernization of the technology there requires to be a corresponding reconstruction (modernization) of buildings. Moral depreciation of buildings with modern productions often come already 10-15 years later and it is very important that the building system has been tailored for the reconstruction and repair. For example, it is known that very often to replace the large-sized equipment in the structures there is no mounting openings, and often almost impossible to qualitatively renew corrosion protection, etc. All this significantly increases operating costs.

There are two strategies in the planning of repair:

1) scheduled preventive maintenance (PM);

2) maintenance of the technical condition.

In the second case, it is assumed that there is a prompt elimination of damage to the extent of their occurrence or replacement of components (including in the framework of the warranty). Such a system requires, in general, slightly lower repair costs, but also requires high-quality supervision of the state of elements of the building. The main task in the system PF - preventing failures is the periodic holding of certain volumes of preventive repair. This ensures trouble-free work of all the elements and the system as a whole. Volumes and terms of works are set depending on the term of service of the specific elements of buildings, and inter repair terms should be appointed before the expiration of their standard terms of service.

Usually the structure has the initial reserve in the form of indicators of operational qualities in relation to the project data of the unification of elements, scheduled backup, etc. The real intensity of degradation (wear and tear), among other things, depends on the initial reserve - the smaller, the more are the stresses in the structure, the higher is the intensity of wear. The strategy of scheduled preventive repair involves periodic restoration of structures till the mass failure (before achievement an inter-repair periods). Failure-free and reliability of the individual elements and the building as a whole has a probabilistic nature, as the place and time of occurrence of malfunctions and failures, the duration of the service to the first refusal, capital repairs, and the subsequent flow of failures and restorations. The total period of service of elements and objects are random functions of time, i.e. performances of a set of simultaneous continuous stationary and non-stationary processes.

The problem of optimization of inter-repair periods with probabilistic positions, especially for the building as a whole, is quite complicated, as in addition to the technical aspect, economic and even social aspects should be taken into account. The probability of the destruction of the bar or the layered system under the same operating conditions depends, in addition, on the number of elements and degree of static indetermination. For these reasons the wider application of monolithic and precast and cast-in-situ structures that have a higher reliability is justified, and in better way will resist progressive destruction. The growth of intensity of a stream of refusals (reduction of the probability of non-failure operation) outside the period of stable (normal) operation in multiple-element system does not mean that it should be stopped. The exploitation of the non-based elements of the building can be continued, as a rule, performing only support repairs or even without repair, causing them for a certain period of work with wear, if it is not connected with a considerable growth of operating costs. Of course, you need more oversight of those elements for the possibility of taking the effective management decisions by the owner of the building.

Already at the design stage it is necessary to decide: either to have big investments in the durability of elements of the building and of the small current costs, or have less initial capital investment in the less durable elements of the building, but the high follow-up costs time and money on their maintenance service and repair. The decision will be taken by the owner of the building taking into account the specific situation, and the degree of inflation, etc. Different systems of approximately equal durability can vary markedly the distribution of failure-free time. In sub-optimal systems the entire service life can be filled with the constant repairs, testifying to the unreliability of the system (building) as a whole. Some experts interpret the standardized service life of an individual item as a term is up to replace it. In this case, the element is subjected to wear without current and capital repairs. However, almost all structural elements and systems of the engineering equipment are being restored, and it is current and capital repairs, combined with partial replacements, ensure their full term of service. The above considerations in one way or another (with account of the level of development of the theory of at that time) were taken into account in the development in the former USSR's system of repair planning.

Sec.11.4. General and Inter-Repair Service Life of Buildings and Structures

The main point in this system is the appointment of common and inter-repair periods of service structures and buildings. In general it is possible to formulate the following definitions of these terms.

• The full codified life of the separate elements and the building as a whole is established by the codes of the total calendar operation time before reaching the limit state characterized by resource refusal. The further operation should be stopped due to unrecoverable violation of safety requirements or because of mass unavoidable «exit» of main structures given by indicators of operational qualities for the minimum acceptable limits which leads to a disproportionate reduction of the effectiveness of operational costs, etc.

• Full actual service life should be (and often is) not less than standardized values, which are installed with a certain confidential probability, and the actual characteristics of the materials hasn't' fully exhausted their resource. The degree of excess of the full actual service life over standardized one depends on the initial designs ' reliability and the level of its technical operation.

• The residual term of service (resource) is total hours from the moment of the survey prior to the limit state.

• The overhaul standard life established in the rules of the approximate average calendar service life between overhauls. It is used for the planning of capital repairs and evaluation of the effectiveness of the organization of technical operation between repairs from the point of view of ensuring trouble-free operation of elements. In this period there is only technical maintenance and current repairs, including supervision of the possible appearance of single failures and evaluating the importance of appearing defects and of the dynamics building technical condition change. To specify, the actual volumes of works on current repair, as well as the dates of the pre-project surveys and capital repairs, depends on the kind of strategy of technical operation is applied at the facility.

At this stage, as a rule, we have to deal with the gradual degradation failures or «crashing» (small failures), not manifested in the form of the emergence of critical defects. Volumes and terms of the restoration measures in this case are governed by the accumulated volume of failures in bearing and non-bearing structures.

When determining the standardized terms of service for buildings, their appointment and the group of solidity was considered. (In [11.18] there are six groups selected for resident buildings based on of solidity, nine groups for public buildings and in [11.19] - seven for production buildings. This data of the documents was preceded by extensive research of depreciation of buildings and their elements in the natural conditions and theoretical study, including probability analysis. Without the analysis of specific numerical data, it should be noted that they were, however, obtained mainly by methods of expert estimates and extrapolations that in this area is very close reception, especially for industrial buildings, where the classification according to the durability is very difficult because of the diversity of functional requirements for the buildings, the types and intensity of the actions in the buildings with various technological regimes. In connection with this, the full period of services for the building as a whole and for its individual elements were standardized for both residential and public buildings, as well as recommended frequency of current and capital repairs, while for production buildings- only periodicity of capital repairs for three groups of conditions of operation was standardized.

Given that mentioned documents have already become a bibliographical rarity, to illustrate the historical approach to the planning of technical exploitation in the table 11.3-11.5 contains fragments of relevant standards related to the basic bearing structures of buildings.

Unfortunately, in the former USSR the system of planning was unsustainable because of the lack of proper control and financing, and in 1991 the provisions of [11.18, 11.19] were abolished, while for the residential and public buildings special document was created [11.22], in which inter-repair terms for elements of the building remained the same, and the table for the standard terms of capital repair of residential and public buildings restated (simplified) with a decrease, in general, inter-repair periods. Total service life of buildings has been excluded from the list of standardized (or at least recommended) settings. I think that this firstly ought to do with, apart from what has already been mentioned, the total period of service is the basis of the standards of depreciation deductions, and secondly, with the the durability of the building as an important parameter when creating a thought-out long-term planning policy.

Table 11.3

Regulatory average length of service of residential houses, their structural elements (From [11.18])

Designation of buildings, their structural elements and decoration	The average length of service in years by groups of solidity of buildings				
	I	II	III	IV	V
1. Residential Houses					
The service life of a residential house in the whole Structural elements of buildings	150	125	100	50	30
2. Foundations					
Strip rubble on a complex or cement mortar, concrete and reinforced concrete	150	125	100		
Strip rubble on lime mortar				50	

178

Designation of buildings, their structural elements and decoration	The average length of service in years by groups of solidity of buildings				
	I	II	III	IV	V
Rubble and concrete pillars					30
Wooden chairs					10
3.Walls					
Especially the capital, a stone (brick in the thickness of 2,5-3,5 brick) and large-block on a complex or cement mortar	143				
Stone ordinary (brick at a thickness of 2,0–2,5 bricks), large-block and large-panel		125			
Stone lightweight walls made of bricks, cinder block and limestone			100		
Wooden log and block				50	
Wooden prefabricated panel, frame: made of clay and adobe					30

Designation of buildings, their structural elements and decoration	The average length of service in years by groups of solidity of buildings				
	I	II	III	IV	V
4.Floors					
Reinforced concrete prefabricated and mono-lithic	150	125	100		
With brick arches or concrete filling on metal beams		125	100		
Wooden on metal beams		80	60		
Wooden on a wooden beams		60	60	50	30
5.Stairs					
Floor-concrete, the level of slab of stone on metal, reinforced concrete string or concrete slab	100	100	100		
Overhead concrete steps with a marble crumb	50	50	50		
Wooden				15	15

Table 11.4

The frequency of selective (SR) and major repair (MR) for the buildings of different groups of solidity (from [11,18])

The name of groups of a building	Total ser-vice life of the building (years)	Periodicity of repairs
A. Residential Houses		
I. Stone building stone, especially the capital one, a foundation of stone and concrete, the walls of stone (brick and large-block), slabs of reinforced concrete	150	SR – after 6 years MR – after 30 years
II. Stone ordinary buildings; foundations are of stone, the walls of stone (brick, large-block and large-panel) floors are reinforced concrete or mixed	125	SR – after 6 years MR – after 30 years
III. Building are from lighter stone; the foundation - stone and concrete, lightweight masonry walls made of bricks, cement and limestone, floor – wooden or reinforced concrete	100	SR – after 6 years MR – after 24 years
IV. The building of wooden are logged and block, mixed; foundations - strip rubble, walls - minced, block, and mixed (brick and wood) , floor - wooden	50	SR – after 6 years MR –after 18 years
V. The building cast-in-frame, frame, Adobe, mud, mud and half-timbered; foundations are on wooden chairs or dummy pillars, walls - timber frame, mud, etc., floors- wooden	30	SR– after 6 years MR– no
Б. Public Buildings		
I. The frame building with reinforced concrete or metal frame, with filling of the frame of stone material	175	SR –after 6 years MR – after 30 years
II. The buildings with stone walls of the single stones or large-block; columns and pillars of reinforced – concrete or brick, floors - reinforced concrete	150	SR– after 6 years MR – after 30 years
III. The buildings with stone walls of the single stones or large-block; columns and pillars of reinforced – concrete or brick, floors – wooden	125	SR – after 6 years MR – after30 years

The name of groups of a building	Total service life of the building (years)	Periodicity of repairs
IV. The building, with the walls of a lightweight masonry; columns and pillars of reinforced concrete or brick, floors-reinforced concrete	100	SR – after 6 years MR– after 30 years
V. The building, with the walls of a lightweight masonry; columns and pillars of brick or wooden, floors-wooden	80	SR – after 6 years MR – after 24 years
VI. The wooden building with round-block or block rubble walls	50	SR – after 6 years MR – after 18 years
VII–IX. The wooden building (frame and panel), tents, trays, stands, etc.	25–10	SR – after 6-5 years MR – no

Table 11.5

The frequency of major overhaul of production buildings depending on the group of solidity and conditions operation (from [11.19])

Solidity of building	The periodicity of the capital overhauls, years		
	in normal conditions	in an aggressive environment and under excessive moistening	during the vibration loads
With a reinforced concrete or metal frame, with filling the frame by stone materials	20	15	6
With stone walls of fat stones or large-block, columns and pillars of reinforced concrete or brick with reinforced concrete floors	15	10	6
The same, with wooden floors	12	10	6
With walls from lightweight masonry, columns and pillars of brick or reinforced concrete, floors-reinforced concrete	12	10	5

Solidity of building	The periodicity of the capital overhauls, years		
	in normal conditions	in an aggressive environment and under excessive moistening	during the vibration loads
With walls from lightweight masonry, columns and pillars of brick or wood, wooden floors	10	7	6
The wooden building with round-block or block rubble walls	10	8	5

In the system of technical regulations of the Belarus Republic in the field of standardization of the technical characteristics of the buildings, methods of their assessment, planning activities for technical operation, etc. [11.23 – 11.28] almost saved specific numerical values of inter-repair periods, in order not to violate the structure of financing if the main assets. There are only slightly modified tables related to the emergence of new types of buildings and types of structural materials [11.24, 11.25]. In fact full service life of buildings indirectly has taken into account the appointment of standards of depreciation deductions (SDD). Data [30], currently being used on the territory of the Republic of Belarus, shows that the existing SDD for similar types of buildings is similar to those adopted in the 60-ies of the past century, i.e., their durability means the same thing. It is obvious that the data in [11.24, 11.25] requires more thorough scientific justification and clarification. The study of the durability of modern building systems with proposals to rationing of their terms of service should go on, considering the foreign experience on similar objects. This gives us an opportunity to develop drafts on a scientific basis, to choose the set of types of structures in buildings from a position of reasonable combination of their durability. In this connection, not quite rational is often used in the present structures, a durable monolithic reinforced concrete frame and the outer walls were made from gas-silicate blocks. Not quite logical, that established in [11.22-11.25] frequency of capital repairs of the individual elements of buildings significantly exceeds inter repair time for buildings as a whole. However, the greatest recommended values, depending on the type of buildings, type of structures, working conditions, etc. will not exceed 30-35 years. I must say that these figures correspond with the real situation of the quality of the construction and operation of buildings in the second half of the twentieth century in the former republics of the USSR.

However, a more than 40-year-old personal practice survey of buildings allows the author to argue that the inter repair time has almost never been observed, the depreciation of structures progressing quickly, and often the question is being asked whether to restore the building in general because it's not only physical repair that is required but also moral one. If the answer of the owner is «no», then it is important to know, what is the remaining service life of the building and what kind of «support» measures should be envisaged to ensure the minimum sufficient safety requirements for this period? If the answer is «yes», then, again, we must evaluate the residual life according to the actual state. Appropriate repair measures should be also developed to set deadlines of their beginning and end, as in many cases the situation is deteriorating rapidly over time. General requirements in this part are presented in [11.23-11.28]. In the context of the above, it should be noted that as stated in the codes of the design – EN-1990 [11.4, 11.29], design working life is essentially inter- repair term, on which the design is to be guided, picking up the parameters that determine the durability of structures (protecting layers and the density of the concrete, etc.), as well as defining the safety settings for the characterization of materials and loads. They are set depending on the class of the building (defined by its purpose, etc.) within the limits of 10-100 years (table. 11.6).

Table 11.6

Indicative value of the project life cycle (from EN 1990-2009)

Category of design service life	Approximate term of service (in years)	Examples
1	10	Temporary structures*
2	From 10 to 25	Replaceable parts of structures- crane runway beams, supports, for example
3	From 15 to 30	Agricultural and similar structures
4	50	The structures of buildings and other conventional structures
5	100	Structures of monumental buildings, bridges and other engineering structures
* Structures or parts of structures, which can be removed for re-use are not considered to be temporary		

Total service life of the buildings in the European standards has not been codified. Therefore when you design a building to the specified in the table 11.6 inter-repair terms, the level of safety shall be appointed through a maximum acceptable probability of destruction, which in probabilistic methods of design is determined by the index of reliability β. Design values β_{tag} are defined in the codes [11.1-11.3, 11.29] depending on the class of the building and the specified service life. Compliance with safety requirements (acceptable probability of occurrence of the limit condition at a given mean time) has ensured compliance with the requirements of standards for the design of structures. For example, for reinforced concrete there is compliance with the requirements of design limit states for the specified index of the reliability, as well as for requirements of the concrete density, protective layers and coatings, requirements of the quality of workmanship and supervision of buildings for specific operation conditions, etc. For exploited structures there are a lot of «inverse problems», for example – presence of residual service life, forecast of reliability for a specified period of operation, assessment of the risks of destruction, etc. Such approach means that from the standpoint of carrying capacity at the end of overhaul period of time (for example, 50 years) when the structure is close to the limit state and for enhance possibility of further operation, it must be restored. From a theoretical point of view this approach is intelligent, because it is obvious that if the design of the structure of the right to full term of service (for example, 150 years) probability of occurrence of the limit state (for a longer period of time) increases. Considering the variability of the loads and the property of the materials, this leads to increasing the required reliability index (in probabilistic design methods) or safety factors (mainly by stress), which results to increasing costs of the materials. However, from a practical point of view, in the absence of rated the end of reference (the full term of service) it is of little use, because it is unclear on exactly how many restore design, what is the outlook of the rating of its durability, especially in conditions of simultaneous exposure to several types of nonstationary, including the aggressive actions, etc. The listing of the equations of the theory of reliability in the general form [11.1, 11.3] for the practical estimation of residual life of structures has not yet been implemented. Possibility of estimation of the full term of service by accurate methods with consideration of periodical repair is questionable. The attempts to solve practical tasks of the assessment and forecast of the state of existing structures in the probabilistic formulation are made only for individual structures of the particular type (bridge, for example) and mode of the actions. [11.36].

For one inter-repair stage of structure and for one type of a load, simplified schematic diagram of the assessment of its technical condition of the account of the probabilistic nature of the resistance of the structure (in the conditions of its degradation, excluding the effect of current repairs) and the variability of the load is shown in Fig.11.3.

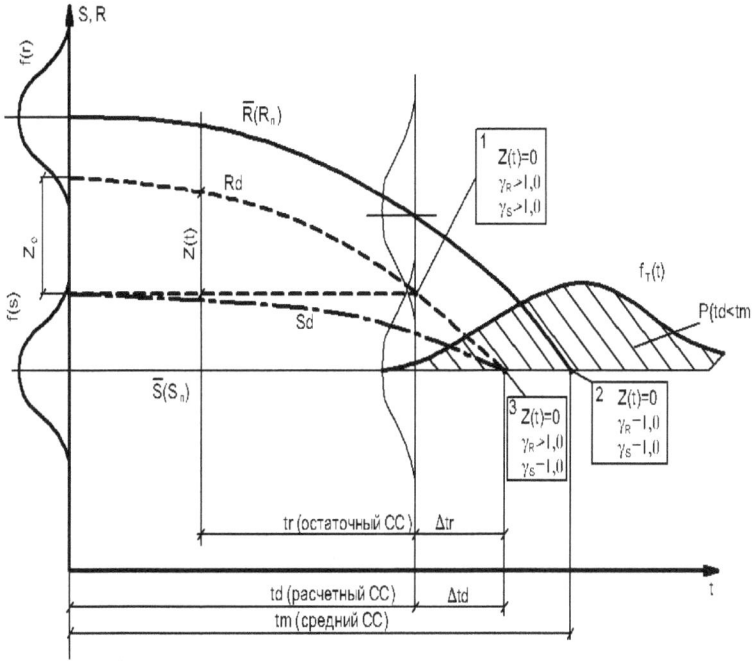

Fig.11.3. Scheme of degradation of the structure to the limit state during one of overhaul period: S – effect of action, and R is the resistance of the structure; $\overline{S}; \overline{R}; S_d; R_d$ -accordingly mathematical expectations and design values of parameters; Z – design reserve of the bearing capacity; t_m – average the(expected) service life; t_d – standardized inter-repair design working-life; t_r – residual service life; $Z(t_1) = 0$ – design limit state, $\gamma_R > 1,0$; $\gamma_S > 1,0$; $Z(t_3) = 0$ – the actual (expected) limit state, $\gamma_R = 1, 0$; $\gamma_S = 1, 0$; $Z(t_2) = 0$ –design limit state considering the reliability load factor $a_s =1,0$. $(\gamma_R > 1)$. Δt_r; Δt_d - a possible increase of estimated working life in the «modification» of load factor

In a deterministic setting, the problems of insufficient reliability of structures at various stages is provided by the excess of the design structural resistance over the estimated stresses of actions, defined at a given level of confidence. The ratio between them (the reserve of the carrying capacity Z) with changes over time, in it's probabilistic nature, makes certain peculiarities and allows in principle and in the semi probabilistic approaches (with the application of safety factors) to take into account some important features of changes of reliability in time. So, obviously, within the limits of the turnaround cycle, properties of the structure degrade, and the statistical variability of properties of materials several increases, as a rule, because of the impact of the additional variability of the parameters of degradation defects. These factors can be measured in situ research in the process of periodic surveys or theoretical calculations, developed well enough for different types of aggressive influences [11.5, 11.31, 11.33].

Stage of completion of the turnaround cycle is characterized by the need of a major overhaul (resource refusal), although in general it can be, as noted earlier, and before it, for example, the accumulation of a certain volume of small failures, etc. (in the system planning preventive repair). Accordingly, the nomenclature of the criteria of fault (the depth of penetration of aggressive actions, the degree of corrosion, etc.) is large enough. There is also quite a large number of methods and techniques of assessment of the time before the corresponding refusal [11.5, 11.31, 11.33, 11.36].

The possible scheme of life assessment of resource failure, caused by violation of the terms of strength, is given in Fig.11.3. Most likely, the average (mean time between failures) term of service is determined by the intersection of the lines of the mathematical expectations of the structure resistance $\bar{R}$ and effects of actions $\bar{S}$ (t_m). It is obvious that in the planning of repairs the relevant terms (design) should be established with a certain probability depending on the responsibilities of the structures, the building as a whole, etc. The design between-repairs period is determined graphically with the cross in time lines of the design resistance of structures and the estimated effect of actions (S_d and R_d), at Fig.11.3. Alternatively, the design lifetime, t_d can be determined by knowing the density function of the probability distribution $P(t_d < t_m)$ on the level of significance $t_d = t_m - \beta_t S_t$. (Some of the documents [11.5] recommend during the assessment of the reliability of existing structures,in a period of time, to take into account the specific residual service life of structures by adjusting values of the indexes of reliability and appropriate safety factors for loads and materials (in deterministic analysis), although the specific methodologies are not been proposed? Apparently, it is understood that, with the passage of time (increasing of the t_i and the reduction of t_r) not only the resistance of cross-section is decreasing, but also the probability of failure in the residual service life, t_r (simultaneous realization of a minimum resistance and maximum load, that has a statistical variability). In other words, for the remaining period you can reduce the required reliability index β_{tag} and, respectively, safety coefficients of materials and loads. In the limit when the $t_i = t_m$ get $\gamma_s = 1$. $\gamma_r = 1$ and is implemented by the refusal of the secondary characteristics of materials and loads (point 2 in Fig.11.3). At that $\beta_t = 0$ and $t_d = t_m$. In the beginning of this chapter there is criticism regarding expression (11.6) for the calculation of R_d, based on the use of the formulas of the so-called «design» characteristics of materials and adopted in the rules of the design of the majority of European countries [11.2].

However, and in relation to the described adjustment of the safety factors (Fig.11.3) at least for the properties of materials (cross-sections), there is also one major objection. The fact is, that the natural variability of the properties of materials has different nature than the variability of loads, and depends on the type and quality of their production, but has no clear mapping to a timeline (remaining period of service) and defined specifically at the time of the survey in accordance with the national standards [11.37]. According to this, in our opinion, the safety factors of the material (or structure in general) should be maintained at the primary level up to the design limit state (point 3 at Fig.11.3). A complete design term of service in this case will increase in comparison with the project for the account of the reduction by Δt_r; Δt_d, but will be less than the average. For the sound of destination and greater differentiation of the current reliability we should conduct more surveys, specifying for each of the (declining) remaining life of the parameters of structures, values β and loads factors, which allows, in principle, specify in the reserve of reliability and delay the exhaustion of the assigned service life.

Thus, probabilistic or semi probabilistic approach allows, in principle, to check the reliability of the structures at the individual inter-repair cycle gradually reducing the level of reliability of the structures which is not below the minimum acceptable value. A further strategy in this approach is not regulated by design standards; the most obvious scheme is as follows (Fig. 11.4).

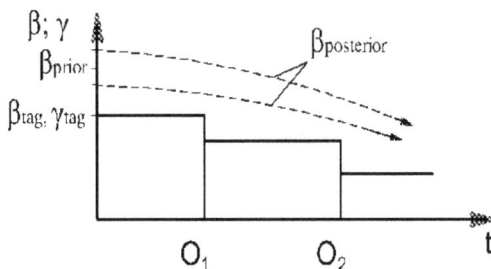

Fig.11.4. The possible scheme of change over time of parameters of reliability.
Q_1 – date of the first survey; βtag - standardized reliability index;
$\beta prior$ – design reliability index; $\beta posterior$ - actual index of reliability

187

After the end of the 1-st of overhaul period the structure will be examined, restored to an acceptable level. The next overhaul period of time is appointed (it may be different from the first depending on the particular situation), respectively, for that a new β_{tag} will be appointed, specify the periodicity of the survey, the specific measures for supervision and maintenance, etc. When implementing such an algorithm it is necessary to ensure the life cycle of the building (service life) , taking into account the following:

The full restoration of structures is hardly expedient, and in our conditions, practically impossible, because of the selective nature of and low quality of the surveys, the presence of hidden defects and low culture of the repair works. Therefore, the intensity of wear on the following post-repair cycle may change significantly (as a rule, in a big way).

2. Depending on the specific type, fullfield work conditions of the structural behavior may change dramatically, may change possibilities of supervision over them, and so on. In particular, the emergence of new finishing (suspended ceiling, decorative panels, panels, etc.) significantly impedes the supervision of structures and at the same time can cause a drop of the temperature-humidity regime in the hidden space, which reduces the durability of structures.

For practical analysis, the fundamental European standards recommended for the existing structures still apply to the whole design lifetime (mean time between failures) for the ultimate limit states the constant value of the index of reliability β equal to the design value for the new (projected) structures β_{tag} [11.13]. Only in individual cases, for structures with a simple calculation scheme and the limited nomenclature of a sufficiently well-studied actions (for example, bridges, where they carry out constant monitoring of their technical condition) contains common suggestions concerning the admissibility of reduction of the required reliability index (safety factors) in time to save resources for the repair works. However, as noted by the leading specialists in this area, when it comes to the security of people, economic problems must become secondary. It is much more important to carefully and regularly carry out diagnostics of structures, to monitor the dynamics of technical condition change. It is known, in general, that much more efficient (up to 3-5 times) funds planned for repairs of short-lived structures invest at a stage of construction in the increase of their initial reliability.

In the Belarus Republic a system of technical regulations has developed, based on previous experience of the former USSR , which regulated the general principles of organizing and conducting supervision of the buildings and structures, rules of their survey and technical operation, and it established the concept and the terminology associated with the types of work carried out in the process of operation of buildings and structures [11.23-11.28]. In these documents, taking into account actual situation in the Belarus Republic minimum requirements established should be observed by the owners of buildings, public and departmental structure. However, till now the general situation with the technical and economic problems in the field of operation of buildings and structures, and also the effectiveness of investments in the sphere of maintenance of real estate is improving too slowly [11.20, 11.21]. It's not possible to continue without a radical change of the situation on the basis of introduction of innovative technologies and methods for the planning of technical operation with the rigid control of the unconditional implementation of the necessary measures.

It is aimed at the recently introduced on the territory of the European Union system of standards ISO 15686 (parts 1-8). «Building and real estate. Planning of service life», developed in the framework of the Technical Committee ISO/TC 59. «Building structures» by specialized Subcommittee SC-14 «Design service life» [11.35]. The main purpose of these documents is a creation of the mechanism of elaboration of rules and measures to guarantee compliance with the requirements of Directive 89/106 EEC and the requirements of the national regulations in the part of provision of security during the remaining period of service of existing buildings and facilities, including optimization of maintenance and repair. It should be noted that many definitions of terms used in these standards, close to the ones used in the technical regulations of the Republic of Belarus. To study the problem it is necessary to introduce a few more useful terms in order.- service life (SL) - the period of time after the erection of the building in which the parameters of operational qualities of the building and its separate elements correspond to the established requirements;

– design life (DL) - is regarded as the alleged specified (appointed in the project) lifetime;

– estimated SL (ESL) - update on field data service life in the concrete conditions of operation;

– reference (RSL)- is established by the corresponding documents of the durability of the standard operating conditions. It can be installed:

a) by the manufacturer,

b) on the basis of previous experience of operation of similar materials or elements,

c) in documents and certificates issued by authorized bodies,

d) in the building regulations

Typical requirement of ISO 15686 is the following - all of the actions for maintenance and replacement of structures, components and materials will be justified economically. In the standard there is a detailed plan of what performance should be considered and evaluated and what performance determined by the indicator of exploitation quality and how (on the whole) they have to be monitored. An important point is the recommended classification of minimal design service life of individual parts of the building depending on the complete design lifetime of the building as a whole (table 11.7).

Table 11.7

The classification of the minimum design duration of service periods of individual elements of the building depending on the full design lifetime of the building

Complex	Design lifetime of the building	Unavailable or the main load-bearing elements of buildings	Elements, the re-placement of which are technically complicated or costly	Mass replaceable elements	Elements and parts replaced in the process of mainten-ance
1	not limited	not limited	100	40	25
2	150	150	100	40	25
3	100	100	100	40	25
4	60	60	60	40	25
5	25	25	25	25	25
6	15	15	15	15	15
7	10	10	10	10	10

Practical recommendations were given for the choice of durability of elements of the building with short or long terms of service. For specific types of elements of the entire mass of self-monitoring you should provide critical, which determine the possibility of the failure, and for them to concentrate efforts on the supervision and recovery. In ISO 15686 is spelled out which actions are the most dangerous for the durability of elements from different materials. The classification of categories of possible consequences of failures and their causes is useful and can be found (table 11.8).

Recommendations are given for selection criteria of failure, requiring major repairs or complete replacement. For the evaluation of the durability of materials and elements of the building in the real conditions of operation and the subsequent planning of their service life, the periodicity of repair and replacement is being by several methods, including the severe, based on surveys, tests (including accelerated) and the study of the intensity of the processes of degradation of structures in the course of technical operation of experienced engineers and experts.

Table 11.8
**Recommended grading the consequences of failure
(extract from [11.35])**

Category	Consequences	Examples of failures
1	Threat to the life	The sudden destruction of the structure
2	The risk of injuries	Defects of stair threads, damage to the stairs
3	Health hazard	The constant dampness
4	Costly repairs	Significant strengthening of the bases and foundations
5	Large costs on multiple repairs	Replacement window shut-off devices, finishing
6	The break in the use of the building	The accident in the system of a heat supply
7	Violation of the level of protection from penetration	Damage of door locks
8	Minor inconvenience	Replacement of lighting devices

It is also possible to have an approximate approach (e.g., factor method), based on the fact that the «standard» durability of the material or element stated by the relevant documents is adjusted with the group of factors, that take into account the specific conditions of operation, the value of which (0,8-1,2) is being established on the basis of expert assessments. Seven factors are recorded:

1) the quality of the components (in the state of delivery);

2) the quality of the design solution (in terms of the availability and reliability of anti-corrosion protection, etc.);

3) the level of performance (in terms of the accuracy of the observance of the established requirements to the quality of construction works);

4) the parameters of the microclimate (in terms of the influence of concrete conditions of operation on the degree of degradation of properties of materials and components);

5) the parameters of natural climatic influences and conditions, such as wind, rainfall, negative temperature and their combinations (in the same part of that and in paragraph 4);

6) the specificity of the concrete conditions of operation, depending on the destination of the building (for example, in residential buildings, at the industrial enterprises);

7) level of service (in terms of accuracy comply with the specific requirements for supervision and maintenance, for example, for hard-to-reach elements, or instructions on the application of the special equipment, etc.).

Already at the design stage (one should provide for the possibility of the application of effective methods of changing the purpose of the building without significant costs), the elimination of the use and dismantling of individual elements and the building as a whole. Possible reasons of such performance include:

– functional (no need for further use of the building, for example, in case of termination of production of these fabrications);

– technological (you need to change the main indicators of a building or items in connection with the change of destination);

– economic (the elements of a fully functional, but their operation costs, for example, replacement of outdated heating appliances to a more effective).

The most effective methods should apply describing interventions in the process of exploitation with minimum economic and social losses. When designing we must bear in mind that the durability of the property (buildings and structures) can be quite large and for the entire service life it can repeatedly change owners, extensions and add-ins and purpose. Therefore, the whole history of its life cycle must be carefully documented. The dismantling of the building with the termination of its use, particularly for buildings, which have not reached the limit of wear, as well as the manufacture of separable and temporary buildings, should be carried out with maximum preservation of structures, assuming their further re-use. (Methods of the decision set out in part 1 of ISO 15686 [11.35] about the main problems arising in the process of operation of buildings, are being specified in the later parts of the standard.

We can conclude in general, that in spite of certain differences of the European codes, connected with the operation of the building, from our one on the composition, content, many terms and definitions, their introduction in the Union of Independent Countries (UIC) as an information document (in the framework of the harmonization process) would use the large experience of the countries of the Euro zone, bring together our codes in the part where it allows the real opportunities and national specifics. In particular, it concerns the system of standards ISO15686-1.

Of course, the best of working abroad in the field under discussion (but not all in a row) should be used in the development of codified and recommendatory documents, be sure to observe the cautious and gradual, carefully study the possible consequences, not to create the contradictions and complexities in the composition of the acting technical and legal documents, and take into account the real possibilities of consumers in their practical use in view of specificity of the present stage. It is obvious that the cancellation of licensing in the field of research, design and production of works in the construction industry with vague prospects for its replacement with the certification and permissions, etc., does not improve at least, the situation with the reliability of buildings and structures, especially in the conditions of the planned transition to the European standards in construction. In the absence of real competition and shortage of finance, while choosing the contractor, the determination (like now) (will be the minimum the declared value of the work, but not the quality.

Ch.12. Operational Reliability of Spatial Hanging Roof

Sventikov A.A.[*]

Sec.12.1. General Comments

The main purpose of the requirements of reliability of buildings is to prevent their failures and collapses. When it comes to reliability of building structures, there is understanding of their qualities to follow out the established or the required indicators in the set limits within the required time [12.1, 12.3, 12.5-12.7].

The probability of non-failure P and the probability of failure or risk $P_f = 1 - P$ [12.5-12.7] are being used as numerical characteristics.

One of the most important issues of safety of buildings and structures is their assessment on the progressive collapse. In the majority of sources this phenomenon is being understood as the spread of the initial local damage in the form of a chain-reaction from element to element, which ultimately lead to the collapse of the whole structure or disproportionately large part of it [12.3,12. 8, 12.10, 12.11]. Note that currently the requirements for the design safety with respect to the spatial bar system are available only in the recommendations «Temporary recommendations on the safety of long-span structures of the progressive collapse in the case of accidental actions», developed in [12.4, 12.8, 12.10]. These regulations provide the development of preventive security measures that reduce the risk of (or probability) accidental actions and size of possible damage.

Existing approaches to assessing the safety of building structures can be divided into the following two classes: direct analysis of the level of risk and the analysis of the stress-strain state of the structures of the system in the destruction of one or several of load-bearing elements (the creation of the possible situation of destruction and analysis of its consequences) [12.11, 12.12]. The last method is widely used for spatial multi-storey reinforced concrete carcasses [12.8, 12.12]. But for the long-span structures, this approach is not applicable in view of the fact that in these systems it is impossible to ensure their functioning at the refusal of the bearing element [12.8, 12.10, 12.11]. Proceeding from this, in the present work probabilistic risk analysis of the progressive collapse of the hanging bar spatial coverings will be performed on the basis of direct assessment of the level of risk on the basis of the preliminary analysis of the stress-strain state of these systems beyond the limit of elasticity.

---

[*] Voronezh state architectural-construction University, Voronezh

The most important task of the analysis of the steel structures is the evaluation of their stress-strain state with the development of plastic deformations. This is done, as a rule, in a deterministic setting without consideration of peculiarities of the statistical distribution of the main parameters of structural systems. In the present work we consider a probability analysis of the hanging bar system with the development of plastic deformations.

Sec.12.2. Analysis of the Hanging Structures beyond Elastic Limit

It is known that the study of stress-strain state of the suspended structures beyond the elastic limit requires to take into account the geometrical non-linearity in addition to the physical non-linearity [12.14, 12.21, 12.22]. Since the hanging structures high-strength materials are mainly used [12.14, 12.15], the relationship between stresses and strains in the material can be written in the form of piecewise-linear functions with hardening [12.16, 12.17]

$$\sigma_i = E \cdot \varepsilon_i \cdot (1 - \alpha_\varepsilon); \qquad (12.1)$$

$$\sigma_i = \sigma_y + E_s \cdot \left(\varepsilon_i - \varepsilon_y \right); \qquad (12.2)$$

$$\sigma_i = \frac{\sigma_t \cdot \varepsilon_u - \sigma_u \cdot \varepsilon_t}{\varepsilon_u - \varepsilon_t} + \frac{\sigma_u - \sigma_t}{\varepsilon_u - \varepsilon_t} \cdot \varepsilon_i, \qquad (12.3)$$

when $\varepsilon_i \le \varepsilon_t$ $\qquad \alpha_\varepsilon = 0$; $\qquad \varepsilon_i > \varepsilon_t$ $\qquad 0 < \alpha_\varepsilon < 1$,

Where E, σ_i, ε_i - modulus of elasticity, the stress and elongation per unit length of the material;

α_ε – factor of the physical nonlinearity (the ratio of the secant modulus of elasticity to the primary) [12.17];

ε_t – elongation per unit length, corresponding the achievement of stresses to the yield strength;

σ_t, σ_u – yield limit and the limit of ultimate strength;

ε_u – elongation per unit length, appropriate for achieving the stress in the material to the limit of ultimate strength;

$E_s = \dfrac{\sigma_u - \sigma_t}{\varepsilon_u - \varepsilon_t}$ - secant modulus of elasticity.

To simplify the analysis of hanging structural systems for physically non-linear analysis we assume that the sagging flexible strings are small and can be neglected [12.14, 12.15]. Then we can write:

$$\varepsilon_t \approx \frac{\Delta_t}{l}; \quad \varepsilon_u \approx \frac{\Delta_u}{l}, \qquad (12.4)$$

where Δ_t, Δ_u - displacements of the end tacks appropriate to the stresses in the material of the thread achieve the yield and ultimate stress accordingly; l – length of the thread.

Given that threads are always in the conditions of the central tension, longitudinal forces in elastic-plastic thread will be determined by the following system of equations:

1) Zone «A» – a zone of structural non-linearity:

$$\Delta \leq 0; \qquad\qquad H = 0 ; \qquad\qquad (12.5)$$

2) Zone «B» – a zone of elastic strains:

$$0 < \Delta \leq \Delta_t; \qquad\qquad H = \frac{E \cdot A}{l} \cdot \Delta ; \qquad\qquad (12.6)$$

3) Zone «C» zone- of plastic deformation:

$$\Delta_t < \Delta \leq \Delta_u; \qquad\qquad H = \frac{E \cdot A}{l} \cdot \Delta \cdot (1 - \alpha_\varepsilon) ; \qquad (12.7)$$

4) Zone «D» - a zone of destruction:

$$\Delta_u < \Delta; \qquad\qquad H = 0 , \qquad\qquad (12.8)$$

where N – thrust in the thread;

EA – axial rigidity of the thread.

There were done test analyses to identify features of the hanging bar structures beyond the elastic limit in a nonlinear formulation of the physical and geometric nonlinearity of the plane lying system with vertical pendants and spatial coverage with the use of double-inclined pendants.

The main parameters of single-span plane hanging structure: span L = 100 m, deflection f = 12,5 m (f/L = 1/8), axial rigidity of the bearing thread EA_k = 12·10^6 kN, flexural rigidity of the beam EJ_b = 2·10^7 kNm, the value of the constant load q = 20 kN/m [12.14, 12.21, 12. 22]. The characteristics of the material of flexible strings were taken based on the recommendations for the design of hanging structures and according to the results of full-scale studies of steel ropes [12.18, 12.19]: ultimate strength – 1400 MPa, the relative yield strength – 750 MPa, elongation per unit length after rupture – 3%, modulus of elasticity of 1.6·105 MPa. The structure was designed on asymmetric temporary uniformly distributed loads value of p. This scheme upload adopted as causing the greatest kinematic movement in the hanging systems [12.14, 12.15, 12.22]. The design scheme and the scheme of upload of the studied hanging system are shown in fig. 12.1a.

To obtain summarizing conclusions of the data by analysis leads to the next dimensionless format:

$$p_0 = p / q; \quad f_0 = f / L. \quad (12.9)$$

where p_0, f_0 the value of the temporary load and the value of the deflection in the relative form.

Deterministic analysis was done by the finite-element method, using a modified method of elastic solutions with the representation of the flexible strings by idealized straight rods considering geometrical and physical nonlinearity [12.21, 12.22].

A plot of the maximum deflections of hanging structure is shown in fig.12.1b when you change the intensity of the temporary load in the design as in the linear formulation and in a physically nonlinear setting. At fig. 12.1c there is a graph of changes of the maximum residual deformations in the assumption of linear unloading of the material after going over the the elastic limit stress. [12.17].

The analysis of the obtained data has shown that in case of the achievementof deflections to values about $1/120 \div 1/130$ L in the rigidity beam of the hanging structure, a significant increase in displacement was observed (see Fig.12.1,b). By the analogy with the plastic deformation of traditional steel beam structures the given state of the hanging structure we have proposed to name the «conditional plastic hinge». Let's note that similar results were obtained for a number of ravines and lattice steel structures [12.13, 12.16], which allow to discuss the possible allocation of a given state to the so-called third limit condition - a state with an unacceptable level of damage [12.7].

In addition, we note, that in the hanging structure with vertical hangs the level of residual plastic deformations, corresponding to the rise of «conditional plastic hinge» $\varepsilon_r = 0.0038$ (see Fig.12.1c), is close enough to the allowable value (0.004) (according to [12.20]), and that confirms adequacy of the methodology of non-linear analysis of the hanging bar systems.

The basic layout settings of the hanging spatial coverage (Fig. 12.2a) are the following: span $L = 72$ m, sag $1/8$ $L = 9$ m, step columns and longitudinal beams – 12 m. The characteristics of the material flexible strings and beam structure were the same as for the flat design. Loading of structure by temporary load carried out according to the asymmetric scheme: in half-span with the uniformly distributed load (Fig.12.2b).

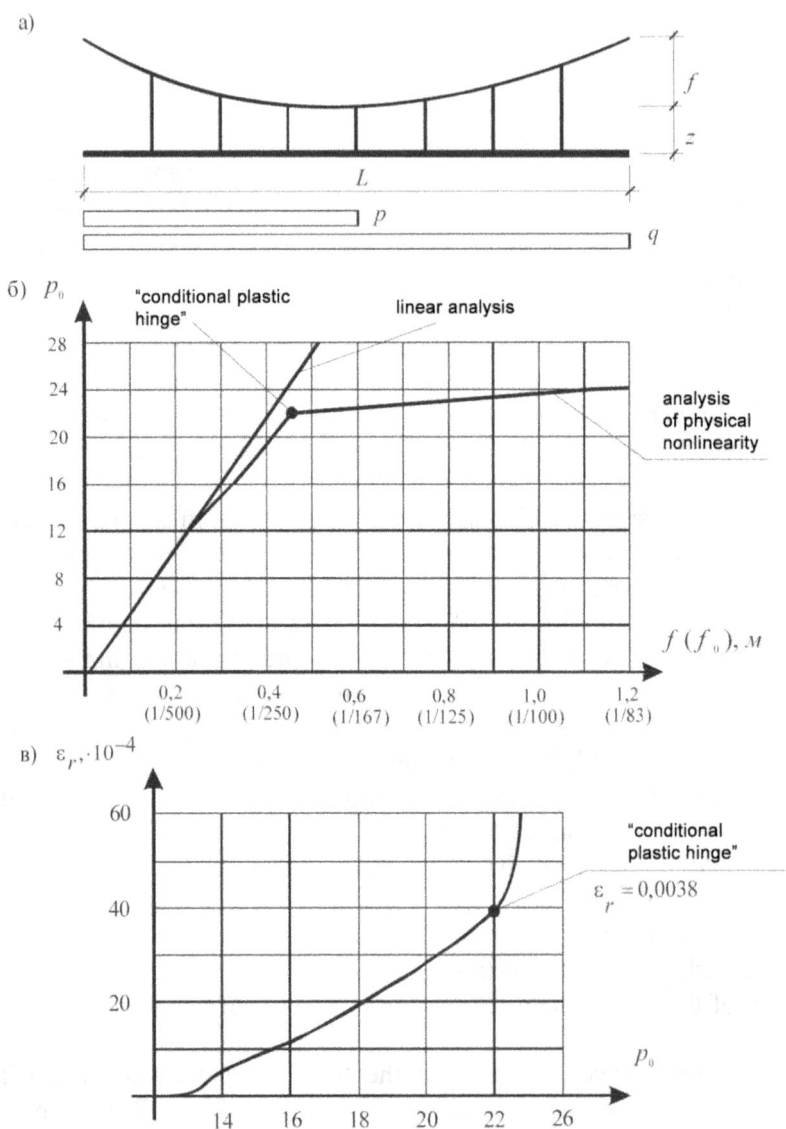

Fig.12.1. Test example of the analysis of the systems with vertical hangs:
a - scheme and scheme upload;
b - graph of changes to maximum deflections of the system;
c - graph of changes of maximum residual deformations

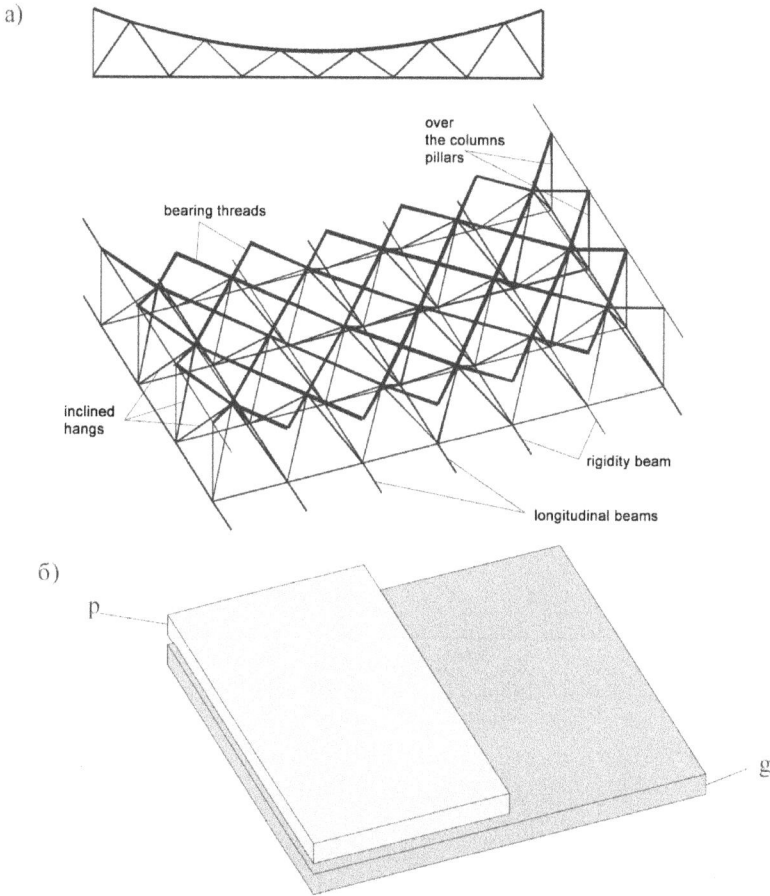

Fig.12.2. Fragment of the spatial hanging cover:
a – structural scheme; b – loading scheme

Figure 12.3 presents similar graphics of the spatial coverage: respectively changes of maximum deflections and changes of maximum residual deformations. From the analysis of the dependencies it is possible to say that for the occurrence of «conditional plastic hinge» for spatial structure happens when the deflection is 1/200 of span. The level of maximum residue of plastic deformations, relevant to "conditional plastic hinge", was more compared by the same value for a flat hanging roof. This fact can be explained with more local character of distribution of plastic deformations and stresses in the rigidity beam in the hanging system with a triangular lattice in comparison with the hanging roof with vertical hangs.

a) P_0

structural nonlinear analysis

32

28

24

20

16

physical nonlinear analysis

"conditional plastic hinge"

12

8

4

$f(f_0), м$

0,2   0,4   0,6   0,8   1,0   1,2
(1/360) (1/180) (1/120) (1/90)   (1/72)   (1/60)

б) $\varepsilon_r, \cdot 10^{-3}$

rigidity beams

80

$\varepsilon_r = 0,0616$

60

longitudinal beams

40

$\varepsilon_r = 0,0321$

"conditional plastic hinge"

20

P_0

16   18   20   22   24

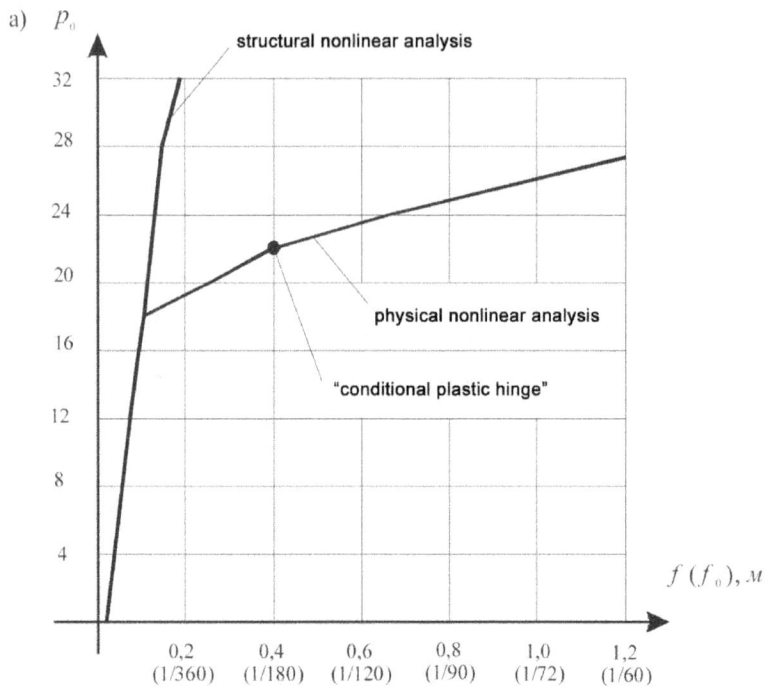

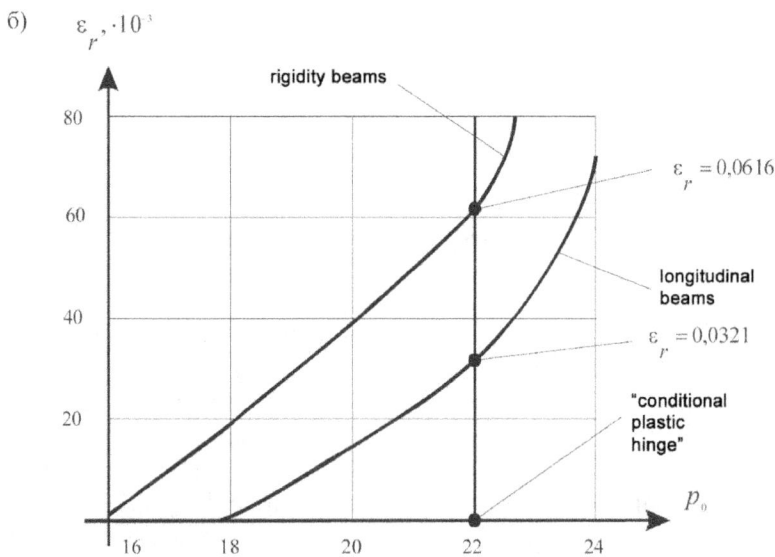

Fig.12.3. The results of the non-linear analysis of hanging space-structure:
a- graph for changes of maximum deflections of the system;
b - graph for changes of maximum residual deformations

200

Sec.12.3 Probabilistic Analysis of Hanging Structures

The probability of non-failure P and the probability of failure or risk $P_f = 1 - P$ [12.5-12.7] has been used as the main quantitative characteristics of the reliability evaluating for building structures. Assuming that the random variables (the strength of the bearing elements of structure and the value of the load) are governed by the normal distribution law, the condition of failure is expressed as follows [6, 7]:

$$\overline{G} = \overline{R} - \overline{Q} \le 0 ; \tag{12.10}$$

$$P_f = \int_{-\infty}^{0} P_G(G) \cdot dG ; \tag{12.11}$$

$$\beta = \frac{G_m}{S_G} = \frac{R_m - Q_m}{\sqrt{S_R^2 + S_Q^2}} ; \tag{12.12}$$

$$P_f = \frac{1}{2} - \hat{O}(\beta), \tag{12.13}$$

where $\overline{R}$ – generalized strength (resistance);

$\overline{Q}$ – generalized load;

$\overline{G}$ – generalized bearing capacity (or reserve of strength);

$P_G(G)$ – the density distribution function of bearing capacity;

Φ – Gauss integral of probability;

β – reliability index (safety characteristic according to Rzanitsyn [12.5, 12.6]).

In view of the fact that the hanging system is quite complex with non-linear character of deformation [12.14, 12.15,12 .21&12.22], the most accurate approach to the assessment of their risk is the use of the static simulation method [12.6, 12.7]. Under this method, the assessment of the probability of failure is carried out by the frequency of events:

$$Q > R . \tag{12.14}$$

Numerical realization of the method consists in the implementation of quite large number of tests according to the Bernoulli, i.e. during each test random realization all baseline values is being generated . After this generation calculation of building structures and checking the condition is determined (12.14). If this condition is true, the outcome of this test is considered to be a refusal. Frequency of occurrence of failure v is considered as the estimated probability of P_f:

$$v = \frac{k}{m} \approx P_f, \qquad (12.15)$$

where k is the number of failures; m is the total number of tests.

As the building structures are the highly reliable systems, it is necessary to study the design cases with a relatively low level of risk of [12.6, 12.7, 12.23], which requires the implementation of a fairly large number of tests (calculations). In the present work to reduce the required number of deterministic recalculations we use the so-called stratified sample [12.6, 12.23]. The initial assessment of the probability of failure was carried out according to the following engineering dependences [12.5, 12.23]:

$$1 - P_{sis} = (1 - P_{thread})(1 - P_{hang})(1 - P_{br}), \qquad (12.16)$$

where $P_{sis}, P_{thread}, P_{hang}, P_{br}$ is the probability of failure of the system (bearing threads, hangs, rigidity beam).

Probabilities $P_{thread}, P_{hang}, P_{br}$ analyzed according to a standard dependencies, respectively, as for systems in series (thread and beams) and parallel (hangers) of the united elements. In the case of a need to clarify the characteristics of the safety β more random generation values of the load and strength have been taken [12.6]. When using the principle of a single destruction, the formula (12.16) is converted into a regular dependence of the estimates of probability of destruction for the in series connected elements [12.5, 12.6]:

$$1 - P_{sis} = \prod_{i=1}^{n_{sis}} \left(1 - P_{f,i}\right), \qquad (12.17)$$

Where s_{is} is the total number of the structural elements.

In the present work, the following conditions of failures are being adopted in the elements of the hanging structure (Fig.12.4):

1. Violation of the conditions of interference stresses elements to the yield limit ($\sigma_i \leq \sigma_t$). This condition will be called "The Failure of Elastic Deformations".

2. Violation of the conditions of interference stress values in the elements according to residual the deformations in case of the plastic hinge ($\sigma_i \leq \sigma_{lim}$). This condition will be called "Failure of Limited Plastic Deformations".

3. Violation of the conditions of interference stresses in the elements of the limit of temporary resistance ($\sigma_i \leq \sigma_u$). This condition will be called "The Failure of the Destruction of the Material."

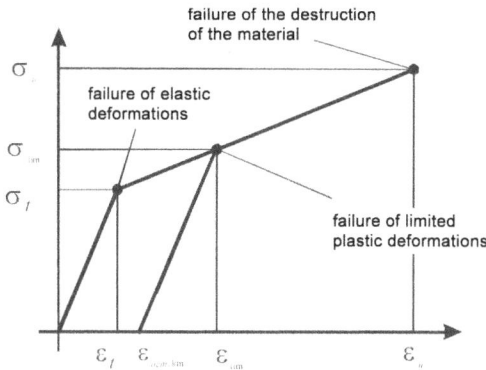

Fig.12.4. Types of failures the material

Note that The Failure by the Criteria of Elastic Deformations is the base for the implementation of the comparisons of the various states of the building structures in view of the fact that the characteristic of safety for different conditions of failures has different values. The second condition is something in between the Failure of the Elastic Deformation and Failure for the Destruction, and will characterize the development of plastic deformations with statistical positions. Fig.12.5 shows the geometric interpretation of statistical factor of the development of plastic deformations. Note that The Failure of the Destruction of the Material is the limiting case of accounting for this factor.

Note that according to the standard (p.1.10 GOST 27751-88) failure of any element of structure can be recognized as an emergency situation [121.8, 12.10, 12.11]. "Any" element of construction is being considered here. Thus, the failure of the destruction of the material in combination with the principle of a single of destruction can be considered a criterion for "the progressive collapse" of a building structure.

Earlier in the analysis of elastic-plastic behavior of hanging structures the position of the "conditional plastic hinge" was identified (a sharp increase of deflections at some stage of deformation). Essentially load, corresponding to the emergence of this phenomenon, is a characteristic of the safe operation of the hanging coverage as a whole. Accordingly, to ensure the operational reliability of the hanging structure the likelihood of settlement of a situation of progressive failure (failure of any element) should not be greater than the probability of a plastic hinge or, in other words, the occurrence of the progressive collapse must be less likely than the rise of plastic hinge in the hanging structure.

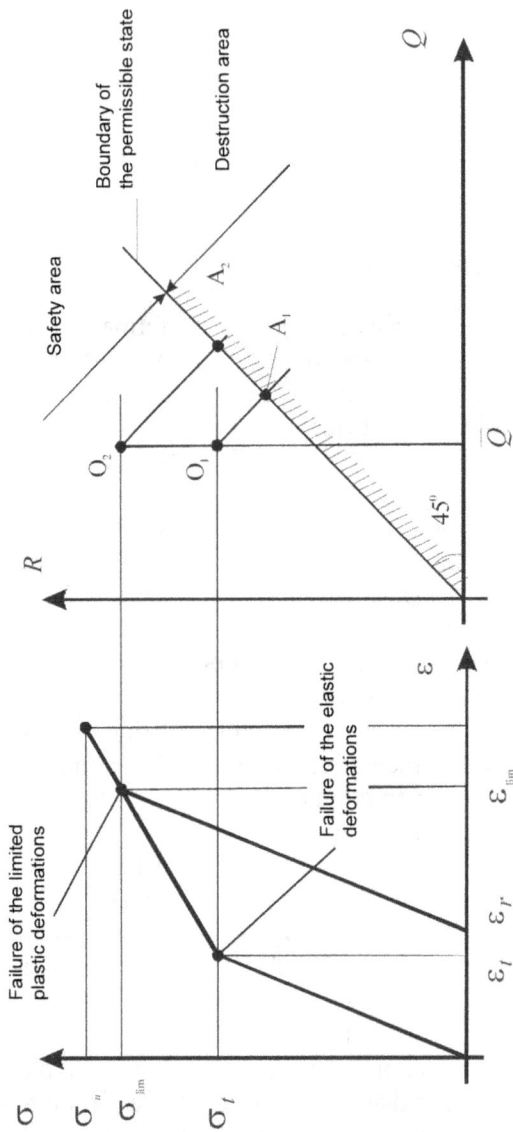

Fig.12.5. Geometrical interpretation of the statistical factor of the development of plastic deformations

Sec.12.4. A Study of the Reliability of the Hanging Structures

The proposed is an estimation of the reliability illustrated by the earlier considered examples of the hanging structures.

It was agreed in the research that a change of a load and strength of elements submits

to the normal distribution law. The scatter of the values of the load was taken 40% ($S_P = 0,4 \cdot \overline{P}$); resistance of steel rolling elements (rigidity beams) - 10% and ropes (bearing strings and hangs) – 7%.

Note that the diagram of the material in case of the failure on the elastic deformations has one straight location and, correspondingly, two defining parameters: σ_t and E. When researching this type of failure we will vary only the limit of elasticity, while elasticity module in all cases is the same. In case of the failure of the limited plastic deformations and destruction deformation diagram of the material consists of two linear locations and, accordingly, depends on four parameters: σ_t, σ_u, E, E_s. It is understood that in the studies of these types of failure only the limit of elasticity varies; the attitude of the temporary resistance to the limit of elasticity, as well as the initial module of elasticity (in the first linear location) and secant modulus of elasticity (on the second linear location) don't change.

Fig.12.6 represents the changes in the characteristics of the safety [12.5] from the relative intensity of the temporary load $P_0 = P / EA$ for these two hanging structures. Note that the criteria of elastic deformations and criteria for the destruction of the material are overall in relation to the criteria of limited plastic deformations and accordingly, we can construct a family of dependencies β, describing one or another degree of development of plastic deformations. It follows, according to these results, that the level of the β for the elastic deformation is about 3.2, and for the limited plastic deformation it is 1.85. The most common standards for the index of reliability [12.7,12. 9] are British standards, according to which, the level of accident risk R or safety characteristics is estimated by the parameter of the population and the type of possible destruction. According to the specified limits of normal conditions of visiting the building codified level of risk has a value of 1.3 -2.5, and for increased conditions- 2.3-3.5 [12.7, 12. 9]. Proceeding from this, we can conclude that our results have a good comparability to these rules and, accordingly, on the admissibility of the proposed probabilistic design provisions the assessment of the reliability of the hanging bar structures.

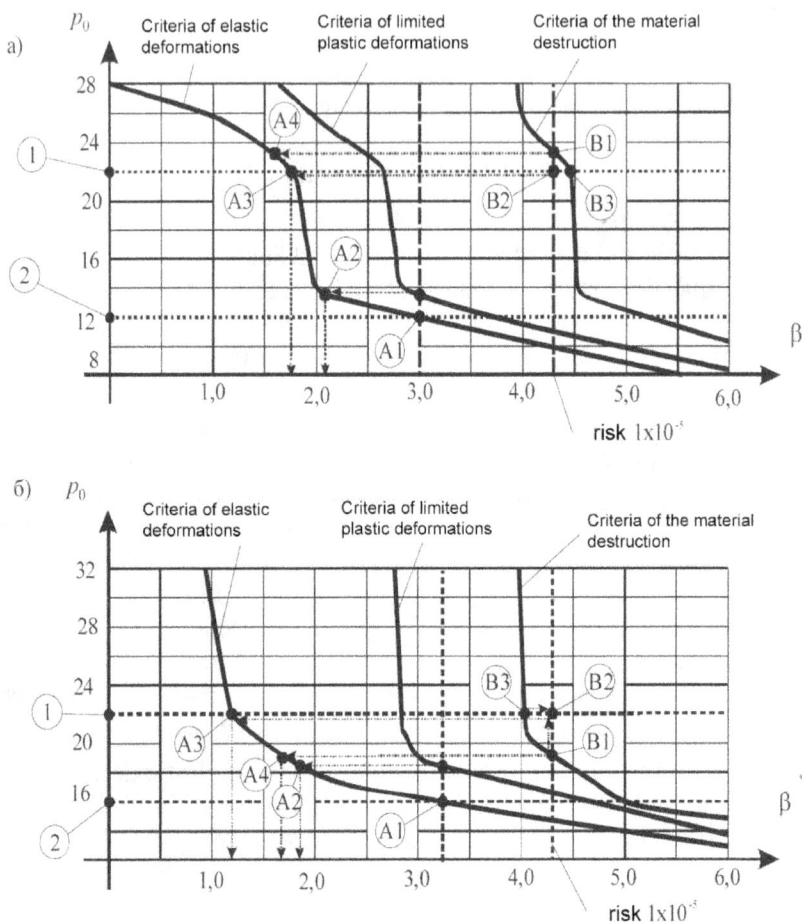

Fig.12.6. The results of the assessment of the reliability of the hanging structures: a - planar design with vertical hangers; b – the spatial structure with sloping hangers: point 1. – the appearance of the plastic hinge; point 2. – the appearance of plastic deformation; A1 – a design point of the elastic deformations. A2 – a design point of the limited plastic deformations; A3 - a design point of the emergence of «conditional plastic hinge»; A4 – a design point of destruction of a material (single destruction of the element); B1 – a design point of the destruction of a material at the level of natural risks (the emergency situation); B2 – a calculated point of a plastic hinge at the natural level of risk; B3 - a design point for the destruction of the material at a level of the load due to appearance of the plastic hinge

206

When analyzing the results, it is accepted that the beginning of plastic deformation (design point A1) is an ultimate condition of elastic deformations, and the estimated point A2 – is the ultimate status to limited permissible plastic deformations. On the basis of the comparison of the safety characteristics in the failure of the elastic deformation (p.A1) and the failure of the limited plastic deformations (p.A2) it is established that the statistical factor of accounting of the development of plastic deformations for the hanging system with a triangular lattice is about 1.2, and for hanging planar structure with vertical hangers is about 1.1. Note that according to the code [12.2] a similar ratio is equals 1.12-1.13, which points to the need to consider this factor for hanging structures with sloping hangers. This fact is explained by the fact that in the hanging structures with sloping hangers the distribution of plastic deformation occurs in a small area of beam rigidity compared with traditional hanging systems with vertical hangers and traditional beaming structures, respectively, that leads to their higher localization and intensity.

Assessment of the probability of occurrence in a situation of progressive collapse should be done by comparing the following design points: B1 - the destruction of any element of the structure in the probability of failure of the natural risk; B2 – the emergence of the plastic hinge in the probability of failure of naturally risk; B3 - the destruction of any element of the structure at the level of the a such load when a plastic hinge in the hanging system will appear.

The analysis of the obtained data showed that in the hanging structure with vertical hangers probability of creating an accidental situation of progressive collapse is less than appearance of the plastic hinge. Thus, we can conclude that for this type of hanging systems factor of the progressive collapse is less significant in comparison with the account of the plastic hinge.

It was also found that for the hanging roof with the inclined hangers under the conditions of equal probability of plastic hinge and the destruction of any element it is necessary to introduce additional safety factor. From a comparison of the characteristics of the safety for the study of the hanging roof this factor is approximately equal to 0.84, there is a sufficient comparability with the requirements of the regulations [12.4]: for the key elements of structures for the buildings with the operation period of over 50 years and a span of more than 50 m $\gamma_{c,add.}$ = 0.8÷of 0.9.

Fig.12.7 shows the comparative graphs of the change of the characteristics of the safety β (analyzed according to the engineering methodology) in groups of elements of hanging structure. Analysis of the results showed that the risk of destruction of hanging coatings is determined primarily by the reliable work of the beam system. This fact is explained by the situation that, as revealed by the analysis of the hanging coatings over the elastic limit, the most «weak» (unreliable) element in them is the rigidity beam [12.21, 12.22]. In carrying threads of the hanging structure the stresses level does not exceed 60% of the limit of elasticity, due to which the a risk of destruction of these elements is significantly lower than that of the entire system as a whole. Also, this fact confirms the admissibility of the adoption to take the destruction of any element of structure as a design situation in the form of the progressive collapse for the hanging covering: at the destruction of any of the girder element, the hanging system may not function.

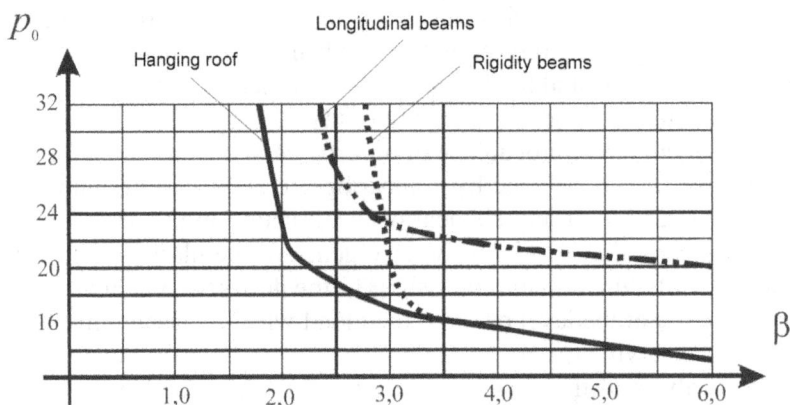

Fig.12.7. Dependence of the characteristics of the safety in the groups of the elements of the hanging roof

As the most important result of this research is the exposure of the reliability of the hanging spatial bar coverings beam elements compared with carrying threads and hangers have the most significant influence.

Ch.13. Creation of Operational Reliability of Buildings as a Complex Systems

Grunin I. YU., Budko V.B., Belykh YU.V.[*]

Sec.13.1. Some Concepts from the Theory of Reliability

The problem of the reliability appeared at all stages of the technological development of the society and stemmed from the practical needs of each man. The questions of the reliability attracted the attention of engineers and have received the relevant decision much earlier of the time, which chronologically is attributed to the emergence of a new scientific-technical discipline – the theory of reliability.

Let us consider a number of provisions and concepts from the theory of reliability.

As in other areas of science, the main concepts of the theory of reliability are perceived by the description of the relations between them. The product means the element, the system or part of it, etc. The quality of the product is a set of properties that determine the degree of fitness of the product for its intended use.

The quality of complex products, such as buildings (or parts thereof), as a rule, is a very large set of properties. With time properties, which are components of the quality of a product, will be adjusted, and, most often, in an undesired direction.

The concept of the reliability is significantly associated with the notion of quality and is one of its properties.

The properties of the determinant of the quality of complex products (the " complex systems"), include:

Safety is the state of the complex system, when the effect of external and internal factors does not lead to a deterioration of the system or to the impossibility of its functioning and development.

Reliability is a property of a complex system which saves in time within the specified values of all the parameters, characterizing the ability to perform the required functions in the given modes and conditions of operation.

The efficiency of the system is the property of a complex system to perform the goal in the specified conditions of use and with a certain quality.

---

[*] LLC «Institute of Technology of Energy Surveys, Diagnostics and Non-destructive testing ",VEMO", Ltd.» Moscow

The majority of technical systems are complex systems, consisting of separate units, parts, components, control systems, etc. Under a complex system we understand an object intended for performing prescribed functions, which can be decomposed into elements (components), each of which also performs certain functions and is in the interaction with other elements of the system.

So for complex systems the reliability is not an exhaustive or dominant characteristic.

Moreover, for some of complex building systems it is impossible, in general, to determine reliability, since many of the positions of the system cannot be in the process of the operation result from the failure or to the efficiency.

Usually for complex construction systems, for example, a large industrial building, one can speak only about the performance or failure for the individual parts, components or systems.

It is necessary to understand, that the reliability theory was not focused initially on the consideration of the systems in their complexity, such as buildings and structures.

Thus, the reliability of the building should be understood as the stability of the quality and efficiency of its functioning, which depends on the reliability of structures and systems of the building in its entirety. Indicator of the reliability of the building as a whole is the optimum time for its trouble-free service.

Analysis of the efficiency of a complex system is connected with the study of the structure and the factors which determine its reliable operation.

Gradual failure is a function of time, which results from the ageing of the material of structures.

Sudden failures are random and are a consequence of the loss of the bearing capacity of structures as a result of stress concentration, exceeding the designed.

In the analysis of the reliability of complex systems they break into subsystems or components, in order to first consider the parameters and characteristics of subsystems (components), and then evaluate the performance of the entire system.

An element integral is a part of a complex system, which can be characterized by independent input and output parameters. In the study of the reliability of the system an element not divided into components and indicators of reliability and durability is considered to be "whole" element. Restoration of the working capacity of an element is possible, regardless of other parts and elements of the system.

Table 13.1

The factors affecting the reliability of constructional objects

Parameter of manageability	Factors
Managed	Construction volume, area of construction
	The use of standard design schemes
	Standardized methods of work
	Technological processes
	The level of specialization and operational division of work
	The number and qualifications of employees
	Composition of construction machinery and equipment
	The duration of works and their complexes
	Availability, capability and efficiency of quality control systems
Partly managed	Technical reasons
	Technological reasons
	Organizational reasons
	Social reasons
	Organization of the construction site
	Availability of material and technical resources
	The availability of energy resources
	Transport links
	Specialization and production capacity of the building organizations
Non-managed	Socio-economic development of the regions
	Natural resources of the region
	Labor resources of the region
	Natural-climatic conditions of the region
	Hazardous geological processes (seismicity, sagging, landslides, etc.)
	Natural disasters
	Other reasons

The event, which contributes to a violation of efficiency of structures, is called a failure. Under failure of bearing and protecting structures we understand the technical position of the element, which precedes the loss of the bearing capacity or full loss of the enclosing function. Consider the classification of failures, in relation to the building constructions (table 13.2).

211

Table 13.2

Classification of failures, in relation to the building structures

Parameter	Characteristic
According to the schemes of the development	In serial
	Gradual
	Sudden
On the scale	Partial associated with the deviation of characteristics of admissible values and do not cause a total loss of working capacity
	Full failures
The degree of significance	Minor, which do not lead to a deterioration in the operational characteristics
	Major (critical), leading to a complete halt of execution functions

Reliability analysis of complex systems has its own specific features. It should be noted that the division of the system for the elements in most practical cases is a conditional procedure and is carried out at a level that is more convenient to consider it. The influence of various failures and decrease of working capacity of system elements differently affect the reliability of the whole system.

In the analysis of reliability of a complex system all its elements and components it is advisable to divide it into the following groups:

1. Elements, the failure of which practically does not influence the performance of the system. Failures (i.e., fault condition) of these elements can be considered in isolation from the system.

2. Elements, the performance of which for the considered period of time practically does not change.

3. Elements, which repair or adjustment is possible when the product or during the stops, and it does not affect its efficiency.

4. Elements, the failure of which leads to failure of the system.

Thus, the review and analysis of reliability are only elements of the latter group. As a rule, there is a limited number of elements, which largely determine the reliability of the product.

These elements and sub-systems are identified when considering the structural scheme of the parametric reliability.

Reliability models establish a connection between subsystems (or elements of the system) and their influence on the work of the whole system. The structural scheme and reliability defines the functional relationship between the work of the subsystems (or elements) in a certain sequence. This scheme is being created based on the principle of the functional purpose of the relevant subsystems (or elements) during the operation of certain part of the work performed by the system as a whole. The technical system can be designed in such a way, that for its successful functioning an efficient work of all its elements is required. In this case it is called a "In Series System". There are also systems in which the failure of one element of the other element is able to perform its functions. Such a system is called "Parallel". Very often the system has the properties of both parallel and in series systems - systems with mixed connection. When analyzing the reliability it is necessary to examine the activity of the system, based on its functional structure and using the likelihood ratio.

Table 13.3

Analysis of positive and negative properties of complex systems, affecting the reliability

Negative	Positive
A large number of elements, the failure each of which may lead to failure of the entire system	Complex systems characterized by self-organization, self-regulation or self-adaptation when the system is able to find the most sustainable state for the functioning
Evaluate the performance of complex systems is very difficult from the point of view of statistical data, as they often are unique or are available in small quantities	For a complex system, it is often possible recovery in part, without interruption of its operation
Even the systems of the same purpose each element has its own minor variation of the properties, which affects the output parameters of the system. The more complex the system, the greater the individual features it has	Not all elements of the system are equally affect the reliability of complex systems

The study of the structure allows you to identify bottlenecks in the design of the system from the point of view of its reliability, but at the stage of designing to develop positive measures to address such bottlenecks. For example, it is possible to analyze how many reserved elements are necessary to ensure a given level of the system reliability. Then you can design the reliability of the system, built from elements with the well-known reliability, or vice versa, on the basis of requirements to the reliability of the system, to present demands to the reliability of the elements. For ease of analysis of building systems there are two possible states: normal operational and failure. In practice, the operation of the housing fund the buildings may have multiple states, corresponding to the partial and gradual failures in the result of accumulation of defects, as well as critical one. In the latter case we have in mind the failures of load-bearing elements of buildings, leading to a total loss of the working capacity of a building.

Assessing the level of operational reliability of buildings shows the correspondence of the status and properties of structural elements to the functional regulations. Therefore, any change in regulatory requirements leads to a decrease or increase of the level of reliability. A typical example of such influence would be changing regulations of thermo technical characteristics of enclose structures. This led to the reduction of operational reliability and a failure for almost all of the buildings built before 1998. A similar situation is observed during the evaluation of the operational reliability of systems of the engineering equipment, power systems, etc. Even at low levels of physical deterioration, such systems do not meet the new regulatory requirements, and are in a state of failure.

From the standpoint of the reliability of a complex system there is a certain dualism, expressed in the manifestation of both negative and positive qualities.

Sec.13.2. Approach to the Creation of Operational Reliability and Safety

During practical realization of works on creation of new generation of systems providing operational reliability and safety of buildings and structures, on the basis of the application of integrated methods of non-destructive control, the necessity arose to develop a unified system of classifiers of integrated safety.

Let's look at a specific example of the general requirements for frame of the monitoring system of the operational reliability and safety of buildings and structures in comparison with the currently existing counterparts (table. 13.4).

The decision of tasks on implementation in the process of creating a system of requirements, discussed above, is carried out on the basis of a complex of the use of means and methods of non-destructive control (table.13. 5).

In fact, there will be created a system for responding to a deviation (hereinafter - event), defined parameters (hereinafter - defects), elements of the subject of the research, with the definition of the reasons of occurrence of defects and means, methods of the response.

Table 13.4

General requirements for the frame of the monitoring system of operational reliability and safety for buildings and structures

Analyzed parameters	Acting analogues	Proposed system
System type	Signal system	The integrated diagnostic system with expert functions, with a dedicated signal block
The current task (total)	Statement of fact of the event The adoption of the decision about evacuation	Prevention of the event on the earliest stages and algorithms prevent events or actions in case of impossibility to prevent
The main principles of the typology of tasks	Salvation	Warning. Salvation. Help
	Realisable direction - safety	Realized directions - safety, reliability, energy efficiency
Main functions	A warning about moving object in the under accidental state	Identification of defects in the early stages of their appearance and development
		Warning of expression and development of the defect. Localization and quantitative estimation of the parameters of defects
		A warning about sudden acceleration of development of the defect with the possibility of projected and actual transfer of the object to the pre-emergency condition
		Emergency response system to any of the events (not accounted for in the development of the system), with the analysis and assessment of the dangers of these events in automatic and semi-automatic mode
	Statement of fact of appearance and development of de-	Definition of the reasons of appearance and development of defects, scenarios of their development and to eliminate in automatic and semi-automatic modes

Analyzed parameters	Acting analogues	Proposed system
	fects. Bound by the attraction of the specialized organizations	The solution of complex tasks of the resource efficiency (including energy efficiency)
The main requirements to the received information	Should be clear to a specialist (for this we introduce support of a specialized organization)	Should be clear to engineers and technical workers of the middle level of the operating organization, directly engaged in monitoring activities
Data collection after the events	After the alarm about the danger data collection stops	After the alarm triggered by event is taking place: – automatic logging of events; – transmission of information in a single duty-traffic control system (SDDS) in the form of a signal about the event; – transmission of information in the SDDS in the form of certificate-memorandum (with indication of the time and place of the event, the history of the development of the defects in the hearth of the events and adjacent areas); – transmission of information in the EDDS in the form of a plan of the proposed action; – automatic connection of reserve capacities of monitoring and control; – external control of the state of the object hardware and non- instrument with harmonized methods
The response of the system of the external factors of special effects to the withdrawal of system failure, unauthorized access	No information	3-stage system of reservation of energy - up to 30%
		The possibility of external interception of control over the system and prevent similar actions of other persons
		The possibility of external influence on the system with a view to its destruction or alteration of disinformation and the prevention of similar actions by other persons

Analyzed parameters	Acting analogues	Proposed system
		The possibility of local inclusion of separate devices and sensors (normally turned off) in order to receive the operative information in a mode on-line
		Protecting the system from any functional change of Control point
		System of access and protection of data
		Remote backup data archiving system
The response of the system to the failure of one or a group of sensors, channel	No information	Connection of reserve energy
		Possibility to obtain the missing data from the failed elements of the system, the design methods for analog statistical charts in automatic mode

Table 13.5

Complex application of means and methods of indestructible control

The basic version of hardware-methodical system of monitoring of operational reliability and safety of buildings and structures
The system is being developed by selection, adaptation under the general model and integration into the overall system of specialized units (stations). In regard to the development of blocks is used the principle of re-use. Development of the system is based on the data of work, the executive documentation, materials of the surveys and ongoing monitoring. The main of them are actual performance parameters. The system is protected from disruptions and a full withdrawal from the system. Permanently damage the system is impossible due to the presence of reserve capacities, as well as reserve hybrid (a combination of hardware and non-instrumental methods) and purely mechanical (non-instrumental methods) subsystems with the possibility to work with the use of paper media. The reliability of the system is achieved by the ability to control the object, even when the complete loss of automated control. This allows you to replace entire assemblies and sub-systems in the course of the operation, without losing control over the installation systems. In connection with the specifics of the means of control, they can also be used as a means of obtaining direct or indirect information in the system of anti-terror.
Block means of the automated control
Station visually-optical and geodetic control The optical system. Optical flaw detectors.

Laser scanning complex. The methods used (indirect visual control):   - reflected and scattered optical radiation;   - spectral method;   - geometric method;   - polarization method;   - holographic;   - observation of double image;   - observation flattened image;   - comparative observations;   and etc., depending on the system configuration.
Hardware-software complex of the thermal control: - thermal equipment;   - contact digital thermometers;   - measuring the density of the heat flow;   - hygrometers;   - anemometers;   - temperature sensors.
Station of ultrasound control: - flaw detector;   - structurescope;   - thickness gauges.   Methods used: - echo;   - echo-shadow;   - echo-mirror;   - shadow;   - mirror-shadow;   resonance and others.
Station of acoustic control: - vibration sensors;   - sensors deformation;   - speed sensors corrosion;   - displacement sensors;   - accelerometer;   Methods used: - acoustic-emissive.
The meteorological station of parametric control: - pressure sensor;   - temperature sensor;   - transmitter;   - sensor precipitation, the direction and strength of wind
Georadar-location complex "LOZA" (without options):   - the unit of probing pulses;

- block of registration; - control and indication unit; - antenna contours; Methods used: - method of subsurface sensing.
Station geotechnical monitoring: Basic set: - accelerometer; - base; - temperature sensors
Control system of engineering equipment: Technological system of collection and processing of information about the state of the engineering equipment, including systems of safety and fire-extinguishing
Block funds mixed (hardware and non-instrumental control)
Mobile hardware-methodical complex service of internal technical control
Block (reserve) of funds non-instrumental control
Reserve complex of non instrumental-methodical and special means for action in situations characterized by the complete failure of all kinds of electronic equipment

Object of research - the complex of buildings and structures is treated as a single object within a complex of natural-technogenic system (hereinafter (СПТS), as the system of interaction of the natural environment and the complex of buildings and structures, providing mutual influence for each other.

Sec.13.3. Methods of the Technical Diagnostic and non-Destructive control

In the course of the development of the monitoring systems of operational reliability and safety of buildings and structures practically solved and justified question of aggregation methods of technical diagnostics and non-destructive control in the system.

Such necessity is caused by the fact that every single method in the theory is incorrect. For example: small signal changes from the objects under study may correspond to large changes in their physical and geometrical parameters.

In other words, in connection with the fact that the effectiveness of any given method turns out to be insufficient, an important problem is the systematical approach to the research of the objects. It practically got reduced to within-methodical integration, based on the use of various physical methods, and among-methodical unification of physical studies, together with other types of research.

Since the objects under study are characterized by a variety of properties and relations, the effectiveness of the determination of the technical condition, after studying them, in the general case, will be the higher, the more extensive there will be a complex. In turn, the increase in the number of in integrated methods leads to rise in price of cost of research and increase the time for their implementation. The problem of the search for a compromise between these factors is one of the most complicated in the theory and practice of integration of the research.

The purpose of aggregation is the choice of such a complex of the methods, which can provide an unambiguous solution of the problem, i.e. a minimum of error in determining the location, the geometry of the system of mutual influence of the investigated objects and accurate transcripts of their physical properties. In determining architecture of complexes one should be guided by certain methodological techniques, i.e. the most rational methods of the work and an interpretation of materials, namely: the implementation of works from the general to the particular, from the study of areas (zoning) to the study of a particular site; from the relatively fast (aerospace, geophysical, etc.) to more detailed field and laboratory methods; by the repetition of surveys of more the accurate equipment on more of a dense network of observations; by the transition from the interpretation of the data of each individual method to the integrated computer processing of all materials; from the qualitative interpretation of materials - to quantification, using the analytical methods of the information processing.

Approach to the unification of methods of technical diagnostics and non-destructive control in the creation of mobile complexes implemented with the establishment of the LLC «Institute of Technology "VEMO" serial line of mobile hardware-technological complexes: "VEMO-2000", "VEMO-building", "VEMO – tunnel", etc.

An example of industrial approach, used in the creation of mobile complexes is shown in Fig.13.1.

In the process of developing information base of the monitoring, there are considerable difficulties with the definition of the basic criteria for the formation of the common libraries, data of the application software.

This is due to:

– the lack of a unified approach to the classification of dangerous phenomena, the ambiguity of terminology and significant readings in the formation of approaches to the study of objects in the construction of the codified base;

– practical absence or extremely blurred definition of the degree of criticality of detected defects, from the point of view of the degree of danger to life and health of citizens;

– notable obsolescence of the codified base in construction and the existence of significant gaps in the system of formation of criteria determine the safety, reliability and energy efficiency;

– noticeable lag of the codified base of the processes of development of the industry of building materials, technical diagnosis, etc.

Creation	Methodical support	Material part of (hardware elements) + delivery vehicles	Software	System of personnel training
Operation	Development of methodical support	Kitting and Assembly	Development and delivery	Training, certification and accreditation
	The expert-consulting support	Service and repair	Updating and consulting support	Training and information support
Modernization	Actualization and renovation	Modernization	Delivery and development of new products	Retraining and re-certification

Fig.13.1. The structure of functioning of mobile complexes

Hardware-software complexes provide a systematic, multi-purpose approach to determine with a high degree of reliability defects, a wide spectrum of characteristics of materials and facilities, appropriate risk assessments and obtain objective results through the integrated application of different methods of non-destructive control and processing of the received multi-parameter information on multitask technology.

As a result of systematization of the received data a conclusion was made about the impossibility to create as a system of clear response on the processes that threaten the health and life of citizens, and an adequate system of evaluation of construction risks, without the development of a unified system of classifiers, implying the possibility of creating in the developed system of internal semantic chain type: environment of formation ↔ dangerous process ↔ object ↔ defect ↔ means and methods of detection ↔ means and methods of response.

Sec.13.4. Analysis of the Problems of Classification

If the quantitative definition of an object is the measurement then the qualitative definition is the recognition and classification. In the process of recognition of the object belongs to some class. This assignment to the class is an analog of attributing the value of the measured properties to some section of the scale for the quantitative measurement. Simply put, a number scaling is intended to be viewed as a special case of the classification. System of classes plays in qualitative methods the same role that the division of the scale in terms of quantity (table 13.6).

Table 13.6

Analyzed parameters	The results of the analysis
Purpose of classification	The purpose of the classification is finding the common properties of the objects. Classifying, we unite in one group of objects that have the same structure or the same behavior
	Classification of claims to the most complete disclosure of qualitative characteristics of an object and is based on the creation of the tree of structural and logical cause-and-effect relations
The limited tasks of the qualitative description	The current practice of deliberately restricts the quality of the description of the purpose of the collection of "grave" characteristics, reflecting the only "important" features of the object. We should not think that such a limitation of the qualitative pattern represents by itself the property of qualitative interpretation. «Rudeness» quality of the picture should be taken as a reflection of the practice of deliberate concentration of analytical tools in the other method of the quantitative description
The limitations of the classification of «border of view»	Any classification is a projection of the introduction of a certain quality «limit of view of» the developer, the relatively limited «quantitative» the measure of view of the object
The conventionality of typing on the basis of «Similar»	Quantitative identification and typing of the status of real-world objects (even with the apparent non-identity states of related objects) defines such status as such on the principle of "the predominance of similarities" (full or partial dominance of these features over the features of the differences). This should take into account the specifics of the selection criteria analysis is usually associated with the risk of «intuitive selection of suitable criteria»

Analyzed parameters	The results of the analysis
Main task	Definition of classes and objects is one of the most difficult tasks of classification.
The impossibility of unambiguous accounting for the majority of determining of the properties	Practically it is impossible to list defining the properties of the object, so that was not an exception. Objects differ in many different parameters. Hence the problem of drawing up of typological series (classifications)
Conventional perception of the impacts of classification models	Mainly in the classifications model of the action «quality change» is a perfect average impact, evenly distributing on all the elements of the object. In fact, the influence is performed on an object only in the limits of the part, which serves as a place of contact, and that is why, in particular, the transformation that is going on in this body, we can understand how "the beginning and the ending". Changing object will always be a heterogeneous environment, which is characterized by a certain spatial scheme of distribution of impacts. In the accordance with the latter body will have the same zones, which has already happened change, as and remaining still in the non-reacted condition. That is, the object that possesses a complex macrostructure, acquires a third-party action only in accordance with a certain sequence of stages
The limitations of the stereotyped perception	Often stereotypes represent the situation at the site locally, without considering the broader picture of the world in which the situation looks just as irrelevant. In connection with this: – "qualities" are often announced it is common signs of characteristics of the object of the research; – the criteria of "quality" is in fact a reflection of the relationship between "better" and "worse"
The limited perception of the «sameness» of the systems and components	Use of the classification models of presentation of the "sameness" of systems means: – firstly, the fact that external influence on the object is interpreted as a conventional "simple sum of" impacts for each of the element; – secondly, it means a loss in our knowledge about the object of the characteristics of the spatial distribution, in accordance with which some elements of the object are defined as "peripheral", when others as "central"

Analyzed parameters	The results of the analysis
	"Sameness elements," ultimately, it should be admitted no more than a reception building homogeneous models.
The problem of terminology	One of the main problems of the classification is the lack of or the variety of options (allowing for different interpretations) the determination of the main determining the properties of objects in the codes
The conflict of classifications (illusion scales)	The conflict of the classifications occurs only in such a case, if the change is imposed and it is on the «randomness». In this connection it is fixed signs of conditions, which include only the differences, which alter the picture of the views of the observed object

Conclusions from the analysis of the problems of classifications can be summarized in the following form:

1. All classifications are of good quality implementations of assessments of quantitative criteria.

2. The classifications are based on the principles of «one tree», are sufficiently robust systems and it is difficult to react to the processes of change (most often in conflict with them).

3. No classification can be able to fully take into account the diversity of determining the properties of an object and can consider the object with only a few (not from all) of the sides.

4. One of the main problems of the classification is the lack of or the variety of options (allowing for different interpretations) the determination of the main determining the properties of objects in the codified documentation.

5. The existence of the different systems of classifications and scales is not a ground for conflict between them.

6. All classifications are inherently uniquely object-oriented, the question of the dominant meaning of one of the classifications of the many classifications describing one object is incorrect, as most of the different classifications, included into the set f for this object consider the subject from different angles of view, based on the different the criteria. It is expedient to consider all of the classifications on the competitive basis, to obtain a more complete picture of the object.

Considering the above requirements was carried out the big work on the creation, the systematization and classification with a view to establishing an information base of safety monitoring.

We will annotate now the main families of classes, the components of the library base of information monitoring of complex safety (without specifying a numerical codes) (table 13.7).

Table 13.7

Code	The name of a family of classes
A	Classifier of the environments of the formation of dangerous processes
B	Classification of hazardous processes
C	Classification of defects of the objects of building and complex engineering objects
D	Classifier of the means and methods of detection of defects of construction and complex engineering objects
E	Classifier of means and methods of response to the manifestation and development of defects of construction and complex engineering objects

Consider a principled approach to the formation of classifiers on the example of the Classifier "A".

The principal scheme of formation of the classifier «A» (Classifier of the sources of danger) is shown in Fig.13.2.

Consider the technology of formation of the Classifier «A» on the example of the fragment of the classifier, which describes the *source of the danger of an explosion in the transportation of explosives* (table 13.8).

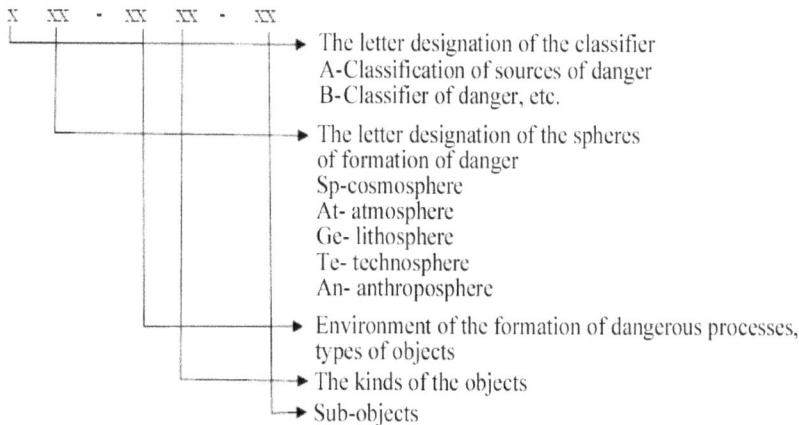

Fig.13.2. Scheme of the classifier "A"

Information disclosed in the table 13.8, can be presented in the form of a coded entry. For example, the source of *the explosion hazard during transportation of explosives with a mass explosion hazard can be described in the form of: ATe-0901-01.*

Table 13.8

Classifier "A" (classification of sources of hazards)

The source of the danger							
The sphere		Environment, type		View		Subview	
Code	Description	Code	Description	Code	Description	Code	Description
Te	technosphere	09	Dangerous goods	01	Explosive materials (EM)	01	Explosives with a mass explosion hazard
						02	Explosive materials, not blowing up in weight
						03	Explosives fire, not blowing up in weight
						04	Explosive materials that do not pose a significant risk
						05	Very insensitive explosive materials
						06	Extremely low-sensitive products

* * *

The practical use of the majority of the elements of the created system of monitoring of complex safety of the construction objects showed outstanding performance and prospects of development of this direction of providing operational reliability and safety of buildings and structures.

The proposed solution doesn't not only contradict but ads, enhances and transfers both of them to providing a qualitatively higher level of security, reliability and energy efficiency of monitored objects.

Ch.14. Safety Provision of the Panel Buildings at the Stage of their Construction

Bayburin A. KH.[*]

Sec.14.1. Methods of Ensuring Structural Safety

Modern regulations require the establishment of new criteria for the quality of construction works based on the methodology of risk and safety analysis of the erected buildings. Under the risk we understand the probability of an accident with the gravity of the consequences. Given that the collapse of structures of residential buildings almost always leads to the death and injury of people, we will consider the risk of accidents with the positions of the probability of collapse (technical risk).

In the analysis of technical risk, we take into account the following assumptions of the theory of the system reliability.

1. An accident of the building comes (with a probability equal to unity) in case of the failure of the foundation or the bearing frame, representing a system within serial connections of elements.

2. The accident of the frame is inevitable if there is destruction of the basic structural cells in the absence of measures against the progressive collapse.

Under the basic structural cell means a volumetric part of a regular structure of the framework, limited by vertical and horizontal load-bearing structures, for example - walls and floors.

The connections of the cells, in this case, can be taken in serial. Otherwise an accident with frame comes with a certain probability, depending on the reliability of the emergency response frame elements (links, joints, assemblies, etc.), connecting the cells and their elements.

3. The accident of the structural cell comes from the necessity of the failure of elements and connections included in the minimum cross-section. From the standpoint of the system reliability due to the connection of components of a cell can be complex, in series-parallel.

4. The collapse of the element of the structural cell comes from the necessity of the failure of all the redundant links and another connection and performing the structure into the mechanism.

Model of the accident panel of the building takes into account these assumptions and is illustrated in the scheme (Fig.14.1).

---

[*] South-Ural State University, Chelyabinsk

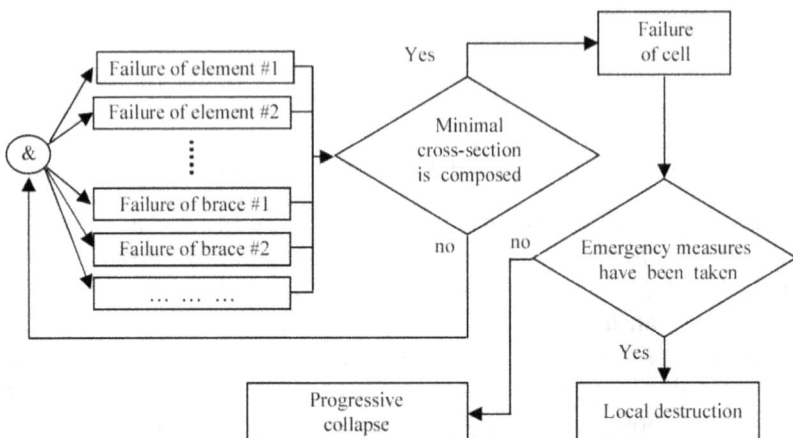

Fig.14.1. Model of the accident of prefabricated building

Consider the methods of ensuring the safety of panel buildings at the level of elements, nodes, links and structural cell.

In the wall of the bearing system panel buildings the platform joint of the support of prefabricated elements is the main node, determining the bearing capacity and safety of structures [14.1, 14.2]. Therefore, for the safety assessment it is proposed to use the indicator of the carrying capacity due to the joint of platform.

$$R_c = R_{bw}t \cdot \left(1 - \frac{(2 - t_m / b_m)t_m / b_m}{1 + 2R_m / B_w}\right) \cdot \left(\frac{b_{pl} - \delta_{pl}}{t}\right)\gamma_{pl}\eta_{pl}, \qquad (14.1)$$

where R – strength of the concrete of the wall on compression;

t – thickness of the walls;

t_m – thickness of mortar joint;

R_m – cube strength of the mortar;

B_w – class on the compression strength of concrete precast element of the wall;

b_m – width of the mortar joint;

b_{pl} – the total width of the platform sites;

δ_{pl} – total offset in the platform junction of floor slabs;

γ_{pl} – factor, taking into account the uneven upload platform sites;

η_{pl} – factor which takes into account the ratio of strength at compression of concrete walls and concrete floors bearing area.

The dependence (14.1) quantitatively expresses the influence of mechanical and geometrical parameters, errors of the device of mortar joints and accepts deviations of the elements on the carrying capacity of the platform of the joint.

In order to eliminate the influence of structural characteristics of nodes panel buildings of various series, we introduce a relative measure of the strength of joints

$$K_R = R_c / R_c^{np}, \qquad (14.2)$$

where R_c, R_c^{np} – the values of actual and design of the bearing capacity.

For the purpose of testing techniques the industrial research of reliability of the construction due to 10-storey panel residential buildings and six houses series 97 and three houses series 121 [14.3] have been carried out. The design of the platform joints of supporting panels of external and internal walls of the house of 97 series is shown in Fig.14.2.

Determining the amount of a controlled sample for each of the studied buildings was carried out according to the following formula:

$$n = \left(t_{1-\alpha} V_R / \varepsilon \right)^2, \qquad (14.3)$$

Where $t_{1-\alpha}$ – a quintile of the t-distribution of the confidence level of $1-\alpha$;

V_R – a variation of the strength of joints;

ε – a relative error.

The minimum volume of the sample at $t_{0.95}$= 1.645, errors ε =0.10 and possible variation of the strength of V_R =0.30, equal to 25. The sample size for each of the buildings was taken equal to 30.

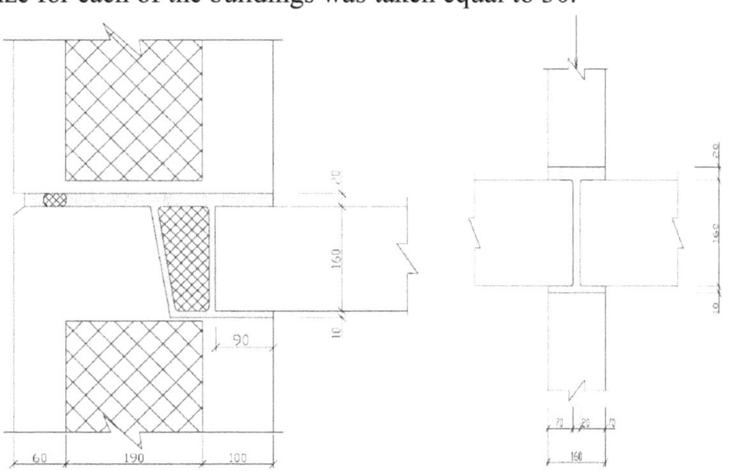

Fig.14.2. Platform joints of the panels bearing areas of the house of 97 series

The relevant united sampling for six houses series 97 and three houses series 121 was taken equal to 180 and 90. Accepted sample provided a high accuracy of the assessment of the design strength of the joints in ε =2.3-3.3% and confidence probability of 0.95.

The histograms of the distributions of the relative index of bearing ability of a platform joints K_R are shown in Fig.14.3.

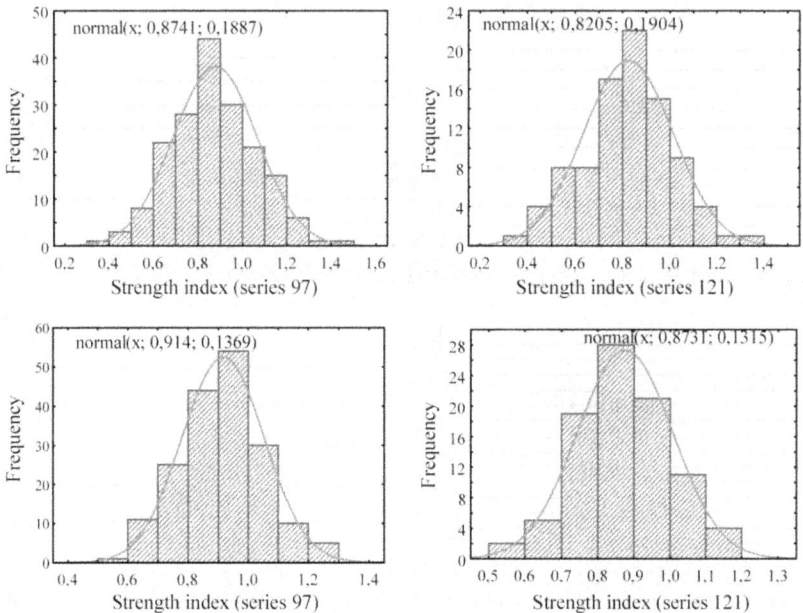

Fig.14.3. The histogram of the distributions of the relative strength index

Analysis of the results showed:

1. approximating the distribution of K_R close to normal;

2. average values K_R less than unit and vary within the limits of 0.821-0.914, which indicates the low availability of the bearing capacity of the joints compared with the average design value;

3. design of standard deviations which are defined by standardized values of the deviations are equal 0.116 for joints of external walls and 0.086 for joints of internal walls and less then observable.

The assessment of normality of received distributions K_R was made by the asymmetry and excess, as well as their standard errors. It is established, that the asymmetry of distributions is close to zero, and the values of the asymmetry and excess lie within its double standard errors. The results of the estimates of approximation by criteria of Kolmogorov and

230

Pearson confirmed the hypothesis of normality of the distributions. The maximum permissible value of the quality indicator of the device of the joint analyzed by the unfavorable combination of marginal standard deviations of [14.4]: from the axes of panels of walls – 8 mm; from the vertical panels of walls - 10 mm; from the symmetry of supporting slabs - 6 mm; the thickness of the panels of walls - ±5 mm; the thickness of the mortar joints – ±5 mm.

With the account of the standard deviations the bearing area in the platform joints of the series 97 may be reduced from 90 to 60 mm in the junction of the exterior walls and from 140 to 110 mm in the junction of the interior walls. The calculated values of the index K_R^{lim} are determined based on adverse of standard deviations and are equal 0.655 for the joints of exterior walls and 0.745 for the joints of interior one. . The probability of the fact that the mean value of K_R will be greater than the maximum permissible K_R^{lim} is

$$\Pr\left(\overline{K}_R \geq K_R^{lim}\right) = 1 - \Phi\left(\frac{K_R^{lim} - \overline{K}_R}{S(K_R)}\right), \qquad (14.4)$$

where Φ – a function of the standard normal distribution, defined by the mathematical table;

$S(K_R)$ – the standard deviation of K_R.

From the observed distributions (see Fig.14.3) it follows- the proba-bility that the values of the quality index of the joints of the exterior walls will not be less than the limiting value $K_R^{lim}=0.655$, is equal to the 97 series 0.8768, to the 121 series 0.8089. In case the strength reduction of the of the joints does not happen (K_R=1), the probability of the observed variance will amount to 97 series 0.9660, for 121 series 0.9653. In design terms of the analysis of the standard deviation for the strength of the joints is 0.116, and the probability of not exceeding with respect to K_R^{lim} is 0.9985.

The probability that the value of indicator of the quality of joints due to interior walls will not be less than the limiting value K_R^{lim} =0.745, equal to the 97series 0.8914, for the series 121 0.8340. In case the reduc-tion of the strength of the joints (K_R=1) does not happen, the probability of the observed variance will amount to the 97series 0.9686 and for the series 121 - 0.9733. In design terms of the analysis of the standard devia-tion for the strength of the joints as well is 0.086, and the probability of not exceeding with respect to K_R^{lim} is 0.9985.

Substantiation of the estimates (basic) values by decrease of the bearing capacity and level of the joints reliability was made with a current analysis of reserves pledged in standards of design of structures. The reserve of the codified material resistance is

$$k_1 = \frac{\overline{R}_b}{R_{bn}} = \frac{1}{1-u_{1-\alpha}V_R},$$ (14.5)

where $\overline{R}_b, R_{bn}$ are the average and the standard value of resistance of the material, respectively;

$u_{1-\alpha}$ is a quintile of the function of the standard normal distribution by level 1-α;

V_R – the coefficient of variation of R_b.

The reserve of the design resistance R_b was determined by taking into account codified values of the of material safety factor γ_m:

$$k_2 = \frac{\overline{R}_b}{R_b} = \frac{\gamma_m}{1-u_{1-\alpha}V_R} = \gamma_m \cdot k_1.$$ (14.6)

In the design regulations for concrete structures the average coefficient of variation adopted for the compression strength V_R=0.135, quintile $u_{1-\alpha}$ =1.645 (if α=0.05 and provision of characteristic resistance is 0.95) and γ_m=1,3. Proceeding from this, the reserve of resistance will be equal to $k_1 = 1/(1-u_{1-\alpha}V_R) = 1/(1-1,64 \cdot 0,135) = 1,28$, and the reserve of the design resistance of concrete $k_2 = \gamma_m \cdot k_1 = 1,3 \cdot 1,28 = 1,66$. Thus, the exhaustion of the reserve is 1.28, which corresponds to a decrease in the strength of approximately on 20%. This result can be taken for the reduction of the relative carrying capacity (border of efficiency). The exhaustion of the reserve is 1.66 and a corresponding decrease in the strength on 40% should be considered as a limit decrease in bearing capacity of a (accidental state of a structure).

Let us find a probability without exceeding design resistance of compressed concrete on the requirements of the code [14.5]. The estimated strength is associated with the average (nominal) strength by the attitude

$$R_b = \overline{R}_b \left(1-u_{1-\alpha}V_R\right)/\gamma_m.$$ (14.7)

The probability that the value of the average strength is greater than the value of the analyzed with the account of (14.7) is

$$Pr\left(\overline{R}_b \geq R_b\right) = 1 - \Phi\left(\frac{R_b - \overline{R}_b}{R_b \cdot V_R}\right) = 1 - \Phi\left(\frac{1-u_{1-\alpha}V_R - \gamma_m}{\gamma_m \cdot V_R}\right)$$ (14.8)

232

The desired probability is equal to

$$\Pr = 1 - \Phi\left(\frac{1-1,645\cdot0,135-1,3}{1,3\cdot0,135}\right) = 1 - \Phi(-2,975) = 0,9985.$$

The obtained value of the probability (without exceeding) will take a minimum, guaranteeing trouble-free condition of the joints of panel buildings. Comparison of the obtained values with the previously analyzed probabilities of ensuring the required quality index showed:

1. The actual error in the mounting of houses provided the conditions $K_R \geq K_R^{\lim}$ with probability of 0.8089-0.9733, which is less than the required level 0.9985; and the probability of failure (risk of accidents) increased in the 18 by127 times;

2. At the observed dispersion of the carrying capacity of the joints, even if the reduction of the strength of the joints does not happen (KR=1), the probability does not reach the required value, that is, there is great variation of the indicator KR.

It has been established that the reduction of the relative bearing capacity of joints for the 97 series is 0.874-0.914, for the 121 series is 0.821-0.873, i.e. no more than on 20%.

To ensure the safety of joints in terms of their bearing capacity, the accuracy of installation should be increased under condition of reduction of systematic deviations KR from the mean values, and under the condition of reduction of influence of random deviation. It is possible to increase the accuracy of mounting in the following ways: to move from a free method of installation to the limited available; increase the accuracy of geodetic breakdown and reconciliation; to introduce a continuous quality control of joints.

Sec.14.2. Risk of Destruction of Prefabricated Buildings

Let us conduct the analysis of accident risk of panel buildings according to the criterion of failure of the most critical components - platform joints of supporting by the logical-graphical method «fault tree» (Fig.14.4). It follows (from the designed tree) that the failure of the platform of the joint can occur during the low strength of the mortar joints, concrete prefabricated elements, increase of the load, reduction of the platforms support (width of the joints) and the increase of the thickness of seams.

With the purpose of definition of priority activities to reduce the risk of failure of the joint one must perform the ranking of factors in order of importance. To do so – examine their influence on the function of the strength and function of the reliability of the joint, that is – conduct a sensi-

233

tivity analysis of the model risk. While calculating the increase in the strength and reliability of a structure with even small changes of parameters, let's determine the weight of type 1 (using the function of strength) and the weight of type 2 (by the function of reliability) of each parameter, as well as the average weight. The results of the calculations are given in table. 14.1.

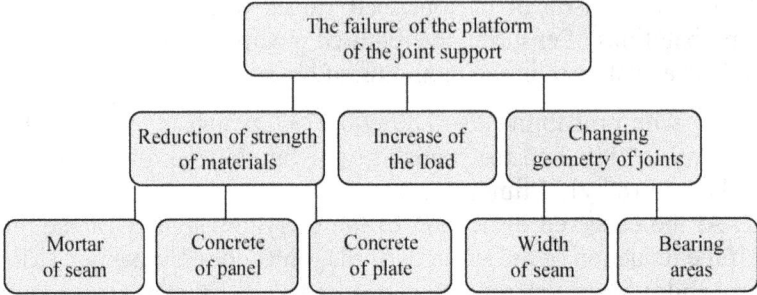

Fig.14.4. Fault tree of the platform joints

The most important parameter is the width of the bearing area, determined by mutual offset bar elements (the weight of 41%). Then the weight of the distributed goes as follows: the strength of concrete panel is 27.9%; the thickness of the mortar joints is 11.2%, the strength of the mortar – 10.4%; the strength of a concrete floor slabs – 9.5%.

Table 14.1

Controlled parameter	Weight		
	type 1	type 2	Average
Strength of concrete of panel, R_{bw}	0,391	0,167	0,279
Strength of mortar of seam, R_m	0,047	0,161	0,104
Strength of concrete of floor plate, R_{bp}	0,039	0,151	0,095
Width of mortar seam, t_m	0,066	0,157	0,112
Width of bearing area of joint, b_{pl}	0,457	0,364	0,410

A list of measures for risk reduction in the order of priorities was made based on the results of the ranking.

1. Ensuring the accuracy of the pair of elements in the interface, through the following measures:

– ensuring the accuracy of the breakdown of axes (the accuracy of the devices, the accuracy of the breakdown, the human factor);

- to ensure the accuracy of elements assembling (the deviation from the vertical and in the plan, the symmetry and depth of support, the installation of the equipment, the human factor);

- ensuring accuracy of the manufacturing of the prefabricated elements.

2. Ensuring the required class of the concrete panels of the walls, depending on the following factors:
– quality of the production of the concrete mix (the composition of concrete, the quality of aggregates, cement and additives);
– temperature-humidity regime for the conditioning products;
– the disassembly and the selling of the concrete strength;
– the accuracy of the control of the strength of concrete;
– lack of a damage during shipment, transportation, storage;
– the human factor.

1. Device uniform seams of nominal thickness:
– the accuracy of the geodetic leveling of the installation of the horizon;
– the device of the beacons by verified thickness;
– accuracy control of the thickness and uniformity of the mortar joints;
– the accuracy of the installation on the elevations;
– the accuracy of the manufacture of the products (the height of the panels of walls, the thickness of plates);
– the human factor.

2. Ensuring the required mark of the mortar seams:
– manufacturing of the quality of the mortar mixture (composition, quality of aggregates, cement and additives);
– ensuring the delivery of the mortar without deterioration of their properties;
– temperature-humidity regime for the soak of mortar;
– time of loading and the strength of the mortar on the each step of loading (especially in winter conditions and when thawing);
– the accuracy of the control of the strength of the mortar;
– the human factor.

3. Ensuring the required class of the concrete floor panels (influencing factors, see p. 2).

As we can see, the human factor is present in every branch of the fault tree. Therefore, the analysis of the potential human and organizational errors, as essential factors of accidents, is an important part of the risk analysis [14.6]. For exceptions (restrictions) of human errors the harmonization of dimensions, reinforcement, strength of elements takes place; the use of the templates, conductors, release tabs on mounting is doing; the use of the standard forms of documents (checklists, charts, schemes) and the reservation of control (a floor-reception, multi-level control) is also doing.

Compiled a list of the risk factors can be considered as a result of identification of hazardous events, which, in turn, can be analyzed by the quantitative method (table.14. 2).

Table 14.2

Risk factor (dangerous event, error)	Probability, P	Weight, V	Risk, PV	Priority
The inaccuracy of a breakdown of axes	0.05	0.410	0.0205	3
The inaccuracy of installation elements	0.30	0.410	0.1230	1
Inaccuracy of control R_{bw}	0.04	0.279	0.0112	4
The inaccuracy of levelling of the installation of the horizon	0.05	0.112	0.0056	5
The inaccuracy of the device seams	0.38	0.112	0.0426	2
Inaccuracy of control R_m	0.02	0.104	0.0021	7
Human factor	0.05	0.104	0.0052	6

The probability of the dangerous events was estimated according to the industrial research of the quality of the erection of prefabricated panel buildings [14.3], and the importance deducted from the table 14.1. The weight of the events can be interpreted as its importance or criticality, and the product of the $P \cdot V$ – risk. By the degree of risk was the ranking of hazardous events.

The methodology proposed in its essence is close to the analysis of species, effects and criticality of the failure, as it assumes possibility and consequences of errors, defects or failures. For comparative evaluation methods analysis of hazardous events is being performed (table 14.3).

In table 14.3 there are critical errors $C_D = D_1 \cdot D_2 \cdot D_3$, where D_1, D_2, D_3 - score estimates, respectively, the frequency, severity and probability of detection of the defect, determined by the special tables [14.7]. After estimating the score, the grouping of dangerous events is being performed in accordance with the frequency-significant matrix depending on the frequency and significance of the errors. Next in the rank of risk set is the status of actions on the possible dangers:

A-high risk; required an in-depth quantitative risk analysis. Causes of errors are subject to unconditional elimination in the workplace and technological design (design change, the use of special equipment, etc.);

B-average risk; desirable quantitative analysis. Causes of errors should be further explored and adopted measures will reduce the risk.

Table 14.3

Risk factor (dangerous event, error)	Score estimates				Rank risk	Priority
	D_1	D_2	D_3	C_D		
The inaccuracy of a breakdown of axes	6	7	5	210	B	3
The inaccuracy of installation elements	8	8	4	256	A	1
Inaccuracy of control R_{bw}	6	6	5	180	B	4
The inaccuracy of levelling of the installation of the horizon	6	6	5	180	B	5
The inaccuracy of the device seams	9	7	4	252	A	2
Inaccuracy of control R_m	5	7	5	175	B	6
Human factor	6	7	4	168	B	7

Priorities coincided with the accuracy of one rank as we can see from the table 14.2 and 14.3. The advantage of the estimation of sensitivity of model risk is a quantitative assessment; the disadvantage is the absence of an assessment of the probability of detection of discrepancies and errors. The limitations of the model of risk should be considered, which does not take into account, for example, risk factors such as the low quality of butt vertical joints grouting, the structure of metal ties, lack of strength of joints and seams in freezing and subsequent thawing, etc. [14.8]. In addition, the failure of the joint can occur when there is a sudden increase in the load in case of accidental actions (explosion, fire, stroke etc.). Given that accidental actions are more likely to receive overhead structures, the safety assessment focuses on the carrying frame. It is almost impossible to determine the failure-free of the structural frame of the building as a whole in view of the complexity of the structural scheme of the reliability. However, given that the panel buildings usually have a regular structure in the plan and in height, there is a basic possibility to isolate some cell of this structure as the basic structural cell. Failure of any of such cells may cause progressive destruction of a building, if there is no a special system of structural relations, individual nodes, elements of connections and joints of panels.

For panel buildings design scheme of the local destruction of accidental actions may be drawn up on the recommendations of the manual on the design of the residential buildings [14.1] and recommendations [14.9]. Neighboring elements influence structures in the zone of destruction, but this effect becomes significant only after their failure. There are two types of structural elements located near the local destruction. In the elements of the first type the impact of the local destruction does not cause a qualitative change of the stress condition and only leads to increase forces and stresses.

In the elements of the second type (this includes structures, that have lost initial support and are located above the local destruction) in the present condition of the building is qualitatively changing the state of stress. The elements of the first type under normal operational actions are subjected to loads in 2-3 times less then destructive, and the local destruction, as a rule, cannot arise the overloading of these structures more than in two times. So, for example, analysis proved [14.3] that in the case of failure of individual wall panels of 10-storey buildings the overload of adjacent panels does not exceed 80-90%, and for 16-floor – 55-63%. Thus, the influence of the neighboring elements of the first type in the design structural cell in the limit of one floor can be neglected.

Sec.14.3. Structural Diagram of the Reliability for Panel Buildings

Based on the assumptions we assume that the safety of the building is determined by the reliability of the basic structural cells, which, in turn, depends on the reliability of the elements and schemes of their connection. As an elementary cell of a panel building it is suggested to consider the four bearing panel walls, covered by the slab floors (Fig.5a). According to the design scheme [14.1, 14.9] the failure of the cell comes with extract of two adjacent panels of walls and at the same time the output of the failure of the vertical links, on which hang upstream panel, or in the destruction of the floor. This scheme corresponds to the destruction of the structural diagram of reliability, shown in Fig.5b.

The proposed structural diagram of reliability can be developed into a more complex one, if we consider the failure of wall panels on the supporting and the average cross sections and the failure of the panel ceilings on bending moment and transverse force, etc. Failures on the deformations and cracks are not being considered, as they are not limited. in accidental situations

The adopted scheme of the system reliability (see Fig.5b) corresponds to the mathematical model

$$P_s = \left[1-(1-P_{14})(1-P_6)\right]P_5, \quad P_{14} = \prod_{ij}^{4}\left[1-(1-P_i)(1-P_j)\right], \quad (14.9)$$

where i. j – deuces adjacent pairs of panels of walls: 1 and 2; 1 and 4; 2 and 3, 3 and 4.

Let, for example $P_1 = \ldots = P_4 = 0.99$, $P_5 = 0.95$ и $P_6 = 0.90$ Then

$$P_s = \left[1-(1-0,9996)(1-0,9)\right]0,95 = 0,95,$$

$$P_{14} = \left[1-(1-0,99)(1-0,99)\right]^4 = 0,9996.$$

Thus, that the reliability of cells practically does not depend on the ties and is determined by the reliability of the floor plate.

In case of overdesign actions (explosion, fire, stroke etc.) panel walls denied with probability 1, that is $P_{14} = 0$. Then, from (14.9) we obtain

$$P_s = \left[1-(1-P_6)\right]P_5 = \left[1-(1-0,9)\right]0,95 = 0,855,$$

The reliability of cells is dependent on the reliability of the system of inter floor relations. If the reliability of connections increased from 0.9 to 0.99, the reliability of the structural cell will rise to 0.941 that is by 10%.

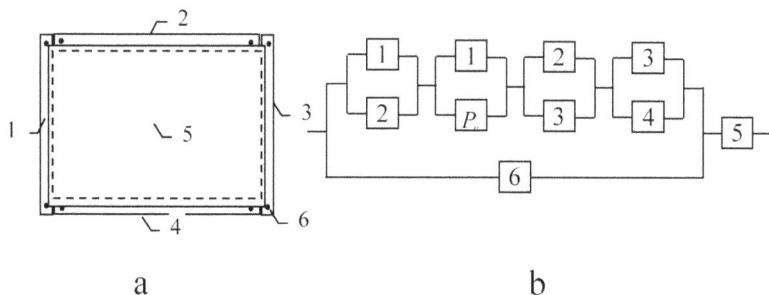

a   b

Fig.14.5. Structural diagram (a) and the design model of the reliability (b) of the elementary cell of panel buildings: 1-4 - wall panel; 5 - floor slab, supported on four sides; 6 - vertical connection between the floors

Resistance to local and progressive destruction is recommended to be provided in the first turn, due to the plasticity of the links between the elements. In the chain «anchor - mortgage - link - mortgage - anchor» on the model of the weakest link the communication is chosen, as puncturing concrete anchoring is fragile, the destruction of the welds – half fragile, and only the link may stretch. Therefore, connection of prefabricated elements, impeding the progressive collapse, should be designed no equal in strength. To ensure that the conditions of the anchor and the weld seams are designed to be 1.5 times stronger, than the link.

Thus, if there are defects in manufacture of the anchor, the mortgage and the device of the welds will lead to decreasing their strength more than 1.5 times, and we will consider that the stability condition to the progressive collapse is violated. The same result would be an increase in resistance of communication in 1.5 times. We classify the possible defects of the device links as following:

1. Manufacturing defects that lead to brittle indent of the concrete of the anchoring zone (change of the size of the anchor, decrease the strength of the concrete in the area of anchorage);

2. Defects leading to the destruction of welds (reducing the size of welds, welding defects and the weld metal);

3. Defects leading to an increase in resistance of communication and brittle fracture at anchor or welded seams (increase in cross-section connection, changing the brand of used steel or properties of the steel welding);

4. Defects leading to a decrease in resistance links below the design value by the condition of the progressive collapse (decrease cross-section connection, undercuts cross-section connection with welding, resistance reduction of steel as a result of replacement or welding).

As the research has shown [14.3], during the erection of prefabricated panel buildings defects of the second class are most likely to happen, that is, defects of the device of welded seams. On the basis of the statistical quality control device seams data we analyze indicators decreasing the strength and reliability of welded joints relations wall panels and floor panels. Strength of welded joints has been tested according to regulations (SNiP II-23-81*). The index of the reliability of the cross-section of metal of the weld and the border fusion is:

$$\beta = \frac{\overline{R}_w - \overline{\sigma}_w}{\sqrt{S(R_w)^2 + S(\sigma_w)^2}}, \qquad (14.10)$$

where $\overline{R}_w, \overline{\sigma}_w$ – the average values of resistance and stress respectively;

$S(R_w), S(\sigma_w)$ – standard deviation of the random variables.

The variability of the stress function $\overline{\sigma}_w = \overline{N}/\overline{k}_f \overline{l}_w$ for the independence of arguments

$$S(\sigma_w) = \sqrt{\left(\frac{\partial \sigma_w}{\partial N}\right)^2 S(N)^2 + \left(\frac{\partial \sigma_w}{\partial k_f}\right)^2 S(k_f)^2 + \left(\frac{\partial \sigma_w}{\partial l_w}\right)^2 S(l_w)^2}, \quad (14.11)$$

where $\overline{N}$ – the average force in the seam;

$\overline{k_f}, \overline{l_w}$ – the average values of the leg and the length of the weld;

$S(N), S(k_f), S(l_w)$ – the standard deviation of the random values.

According to [14.1] the metal connections between the inner and outer panels are analyzed on the perception of the force of the tear-off position and are not less than 1tn for 1 m of the outer wall along the facade within a height of one floor. Communication floors should take tensile efforts along the length of the building – 1.5tn for 1 m width of the building, and across the length – 1tn on the 1 m length of the building.

The design length of the bilateral welds is equal to 40 mm, single-sided – 80 mm, width of seams - 6 mm. The transition from a width to a leg of the seam was carried out according to the formula $k_f = 0,5(b_w - 1)$. For the design conditions the following values of the variability of the random variables were used and they were obtained from the tolerances specified in the standards: $S(k_f)$ =0.07 sm, $S(l_w)$ = 0.2 sm. The variation of the resistance of the metal was taken equal to 8%, the load – 20%.

Design reliability of welded joints of metal of the weld and the border fusion – 0.992-0.999. The results of the calculations for the links wall panels are summarized in table. 14.4. This table shows the greatest values of the decrease of the strength and the reliability for the design of metal of the weld and the border of fusion on the installation of six buildings.

Table 14.6

№ buil ding	$\overline{b_w}$, cm	$S(b_w)$, cm	$\overline{l_w}$, cm	$S(l_w)$, cm	K_R	P	K_P	K_Q
1	3,42	0,91	5,10	0,15	0,701	0,716	0,732	26,50
2	3,40	0,84	5,90	0,10	0,833	0,864	0,883	10,32
3	4,35	0,75	7,70	0,13	1,457	0,954	0,976	3,73
4	4,58	2,16	6,90	0,11	0,978	0,881	0,901	11,49
5	7,63	0,44	8,00	0,08	1,090	0,997	0,999	3,36
6	7,89	1,11	6,80	0,16	0,934	0,875	0,876	15,70
Note. R – reliability (probability of failure); K_R – an indicator of the strength reduction; K_P is the factor of reduction of reliability; K_Q – an indicator of the increase of probability of failure								

Deviations sizes of the welds relations of wall panels lead to a decrease in the strength of a seam at 2-30%, reducing the reliability of the 2–27% and increase the probability of failure of the connection in 4–27

241

times (Fig.14.6). Although defects in welds do not lead to reduction of their strength more than 1.5 times, the condition of stability to a progressive collapse is being violated, because the link will be destroyed half fragile on the weld.

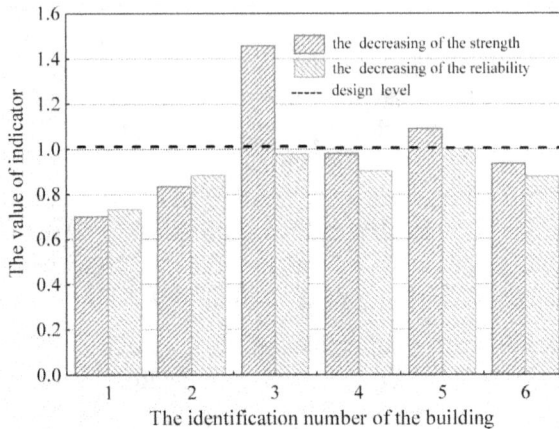

Fig.14.6. The influence of the defects on the strength and reliability of welded joints of wall panels

In the compounds of slabs the actual strength of welded joints corresponds to the design values (without verification of the physical-mechanical characteristics of metal and internal defects), and the reliability of the joints is reduced insignificantly by 0.5–8%. This is due to more convenient for welding location of connections or links.

Thus, the probability of progressing collapse of panel buildings in case of accidental actions is increased many times over, in the observed defects of the device connections and welds. To remedy the situation there is a need for increasing of technological discipline, including all types of control and supervision over the quality of installation of prefabricated panel buildings.

* * *

On the basis of the formulated assumptions of the theory of system reliability with respect to the structural accidents a model proposed of an accident panel buildings. It is established, that the methods of ensuring the safety of the buildings to be considered at the levels of a bearing frame, structural cells, elements, nodes and links.

It is invited to adopt the compression resistance of platform of the joint as a measure of safety of panel buildings at the stage of their assembly. The relative value of this indicator K_R expresses decrease in bearing ability and reliability of joints during admitted defects, detected in the statistical quality control. It is proved that the random values of the index K_R are normally distributed. The limit values of the indicator and the availability of the random values are being justified, allowing with the minimum expenses perform quality control of installation of buildings on the criteria of safety of the most critical components.

A complex of measures for reduction of accident risk was proposed on the basis of the tree of the failure of joints, as well as the method allowing you to rank the priority of these activities for a quantitative estimation of the degree of risk.

A model for the analysis of the system reliability of structural cells was also proposed on the basis of the analysis of the design schemes due to the progressive collapse of panel buildings, which determines the safety of the building as a whole. It is shown that the risk of collapse in case of accidental actions depends, mainly, on the reliability of the system of relations. The possible defects of relations, which may lead to violation of conditions of stability to a progressive collapse were analyzed. Operational research has established that the observed defects of the device welds of the links lead to a significant reduction in the level of safety of buildings. The use of the proposed approach allows to evaluate the safety of panel buildings in accordance with the requirements of technical regulations «On safety of buildings and structures».

Ch.15. Evaluations of Existing Structures

Raizer V.D.[*]

Sec.15.1. Principles of Assessment

Assessing the reliability of an existing structure may however differ very much from that for a structure at the design stage. The existing structures are subject to deterioration and damage, they are likely to be periodically inspected and repaired or strengthened if necessary. Initially, the various uncertainties related to loads and resistance parameters were assessed based on the pre-existing data. Later on the data can be updated based on actual observations with the structure in service. Consequently, regarding the state of information the situation in assessing the existing structures is completely different from that during design. Moreover, special attention is paid to specific parts of the existing structure, especially to those that are at high risk of damage according to observations of the structure in service [15.1]. On the other hand, the interpretation and analysis of additional information may not be simple. The assessment of actual reliability of the existing structure will includes one or more of the following actions.

• Modification of the existing facility to add new load-bearing structural members to the existing supporting system.

• Adequacy checking to determine whether the existing structure can resist loads associated with the anticipated changes in operation or service life extension.

• Repair of the existing structure, which has deteriorated due to time–dependent environmental effects or damage from accidental actions.

• Checking if there are doubts about reliability or structural integrity of the structure.

The analysis and design for assessing the existing structure shall be based on the general principles. A better judgment of the structure can be made on the basis of quantitative inspection. These should result in a set of values, which characterize the current properties of structural elements. To perform inspection one should have information on the probability of detecting some damage if present and the accuracy of the results [15.2]. The inspection can include a specific investigation of the standard design load. Based on the tests, one may draw conclusions with respect to:

• load- bearing capacity of the member in question;
• its connection with and influence by other members of the structure;
• actions of previously unaccounted probabilistic loads;
• behavior of the whole structure under load.

---

[*] Reliability Engineering Consulting, San Diego, USA

The inference in the first case is relatively easy – the probability density function of load-bearing capacity is simply a cut-off at the value of the proof load. The inference for other conclusions is more complex. The number of proof load tests should not be restricted to one. The tests can be performed on a single member (or a sample of structural elements) under various loading conditions. To avoid unnecessary damage to the structure, the proof load should be applied in gradually increasing increments with recording the structural response (stresses, deformations, etc.). The measurements may also give a better insight into behavior of the system. The standard design loads can sometimes ignore the operational wear. However these factors should be taken into consideration in design. Based on the investigation results, the structure or its parts can be modified and its reliability re-assessed. To do so, two different venues can be explored:

• correcting the multivariate probability distributions of separate variables; this method can be used to determine the updated design values to be used in the partial factors format and for comparing action effects directly with the limit criteria (cracks, displacements);

• formally updating the structural failure probability.

In principle the result of all observations (qualitative and quantitative inspections, calculations, proof loading) should be processed in either of the two venues.

It is always useful to make also an initial visual inspection of the structure to feel its condition. Major defects could be evident to an experienced eye. In case of severe damage, immediate measures (including abandonment of the structure) may be taken. To better understand the present condition of the structure, one should simulate damage to the structure model and estimate intensity of various loads and/or physical/chemical actions. If there is a discrepancy between calculations and observations, it might worthwhile to look for errors in design and construction, or load assessments.

Sec.15.2. Reliability Assessment

Having considered a structure in its current state and using the current information, reliability of the structure is estimated by means of failure probability. It can be noted that the model of the present structure may be different from the original model; if the reliability continues to comply with structural codes and design specification, no further action is required.

When only a single load is applied to the structure, failure occurs if and only if $r < f$; where $\tilde{r} = \tilde{R} / R_d, \tilde{f} = \tilde{F} / F_d$, are dimensionless values of random resistance and load, $\tilde{R}, \tilde{F}$, respectively; R_d, F_d, are their design values. It is assumed that r and f are independent random variables with log-normal distributions. Then, reliability index becomes [15.3]

$$\beta = \frac{\overline{\rho} - \overline{\omega}}{\sqrt{s_\rho^2 + s_\omega^2}} \qquad (15.1)$$

where $\overline{\rho} = \ln \overline{r}, \overline{\omega} = \ln \overline{f}$. Let's assume that a structural element is subjected to a proof load increasing up to the value of $f = f_0$ and that the element sustains that load. (Subsequent analysis is after [15.4]). If f_0 is fixed with insignificant uncertainty, it can be concluded that ρ has the conditional distribution function

$$P_\rho(x) = \Phi_\gamma \left(\frac{x - \overline{\rho}}{s_\rho} \right), \qquad (15.2)$$

where $\gamma = \dfrac{\ln f_0 - \overline{\rho}}{s_\rho}$ \qquad (15.3)

$\Phi(\gamma)$ is the lower truncated standardized normal distribution with truncation point γ.

The conditional failure probability then is Given the structure in its present state and given the present information, the reliability of the structure is estimated, by means of a failure probability. It can be noted that the model of the present structure may be different from the original model; if the reliability is sufficient (i.e. better than normally accepted in design) one might be satisfied and no further action is required.

In the case when only one load is applied to the structure, failure occurs if and only if $r < f$, where $\tilde{r} = \tilde{R} / R_d, \tilde{f} = \tilde{F} / F_d$ are dimensionless values of random resistance and load $\tilde{R}, \tilde{F}$ correspondingly; R_d, F_d are their design values. It is assumed that r and f are mutual independent logarithmic normal random variables and reliability index becomes [15.3]

$$\beta = \frac{\overline{\rho} - \overline{\omega}}{\sqrt{s_\rho^2 + s_\omega^2}}, \qquad (15.1)$$

where $\overline{\rho} = \ln \overline{r}, \overline{\omega} = \ln \overline{f}$. Assume that the structural element is subjected to a proof loading up to the value $f = f_0$ and that the element survive that load. (In the subsequent analysis we follow work by [15.4]). If f_0 is fixed without significant uncertainty, it can be concluded that ρ has the conditional distribution function

$$P_\rho(x) = \Phi_\gamma \left(\frac{x - \overline{\rho}}{s_\rho} \right) \tag{15.2}$$

where $\gamma = \dfrac{\ln f_0 - \overline{\rho}}{s_\rho}$ $\qquad\qquad$ (15.3)

$\Phi_\gamma(\cdot)$ is the lower truncated standardized normal distribution with truncation point γ. The conditional failure probability then is

$$P_f(r < f) = \frac{1}{s_\omega} \int_{-\infty}^{\infty} \Phi_\gamma \left(\frac{x - \overline{\rho}}{s_\rho} \right) \varphi \left(\frac{x - \overline{\omega}}{s_\omega} \right) dx \tag{15.4}$$

where $\varphi(x) = d\Phi(x) / dx$

Example: *For the values*

$s_\rho \approx v_r = 0.2, s_\omega \approx v_f = 0.5, \overline{\rho} = 4\sqrt{0.29} = 2.154, \overline{\omega} = 0$ *the unconditional reliability index β gets the value of 4. The conditional reliability index corresponding to Eq.(15.4) is shown in Fig.15.1. It is seen that the proof loading must be made at rather high levels relative to the strength distribution in order to obtain that an essential increase of the reliability index value above 4. For example, only an increase of 5% (from 4 to 4.2) is obtained for $\gamma \approx -1.3$. This fractile value corresponds to about 10% occurrence probability of failure during the proof loading.*

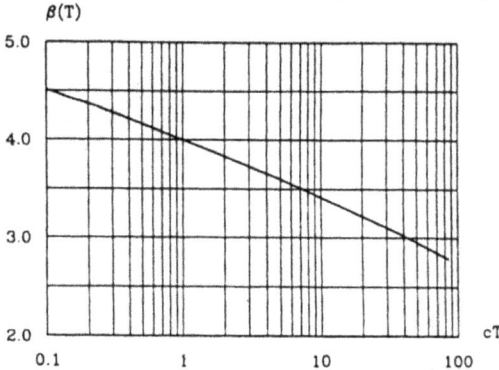

Fig.15.1. Conditional reliability index β (γ) as a function of the proof load level γ

A situation can also be considered when the structural element is subject not only to the effect of a single load, f, but to a sequence f_1, $f_2, ..., f_m, ...$ of mutually independent effects with identical logarithmic normal distributions. Moreover let us assume that it was erroneously taken in the design stage that the element would be subject to one and only

247

one such load effect. The element is installed in a structure for which the total service life was planned to be T. The loads effects $f_1, f_2, ...$ have short durations and occur in succession but randomly in time, implying that there is random number $N(\eta)$ of them within a time interval of given duration η. It is assumed that $N(\eta)$ has a Poisson distribution with parameter $c\eta$ where c is a constant, which can be interpreted as the mean value of the number of load effects per time unit. Thus its probability is [15.5]:

$$P[N(\tau) = n] = \frac{(c\tau)^n}{n!} \exp(-c\tau).$$ (15.5)

Given that $N (T) = n$, the maximum value of $lnf_1, ..., lnf_n$ gives the distribution function $\Phi\left[(x - \overline{\omega}) / s_\omega\right]^n$. The distribution function of

$$lnf = \max \{lnf_1, ..., lnf_{N(\eta)}\}$$ (15.6)

can be obtained by the use of (15.6) as

$$P_{lnf}(x) = \sum_{n=0}^{\infty} \Phi\left(\frac{x - \overline{\omega}}{s_\omega}\right)^n \frac{(cT)^n}{n!} = \exp\left[-cT\Phi\left(-\frac{x - \overline{\omega}}{s_\omega}\right)\right].$$ (15.7)

An elegant way to see this result is as follows. If only those points of the Poisson process of intensity c for which $lnf_i > x$ are considered, then a so-called thinned Poisson process of intensity $cP [lnf_i > x]$ is used. Obviously the event $lnf \le x$ occurs if there are no points from thinned Poisson process during the time T. The right side of (15.7) is then directly obtained as the occurrence probability of this event corresponding to $n = 0$ in Eq. (15.5). At $cT = 1$ and for large x the distribution function (15.7) approximately equals to $\Phi\left[(x - \overline{\omega}) / s_\omega\right]$. In particular at $cT = 1$ the effect of the simplification used in the design phase is therefore without importance. The correct reliability index is

$$\beta(T) = \Phi^{-1}\left[\int_{-\infty}^{\infty} \exp\left\{-cT\Phi\left(-\frac{x - \overline{\omega}}{s_\omega}\right)\right\} \phi\left(\frac{x - \overline{\rho}}{s_\rho}\right) \frac{dx}{s_\rho}\right].$$ (15.8)

This reliability index for a structural element with logarithmic normal strength is shown in Fig.15.2 as a function of the mean number cT (for the parameter values considered in previous example) of mutual independent identically distributed load effects within the time period T. Parameter c is the intensity of the occurrence of the load effects in time by the Poisson process.

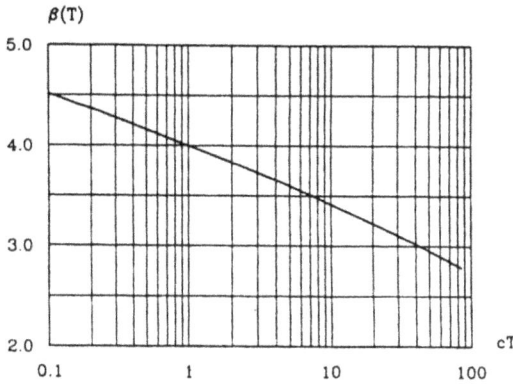

Fig.15.2. Reliability index β *(T)* as a function of the mean number cT

Due to a repair situation at time t after a period of operation of the structure it is found in the original analysis that after reaching cT the limit load, instead of being close to the unity is actually close to about 100 (i.e. highly unsafe). Therefore a decision is made to re-evaluate the reliability of the structural element with the respect to failure during the rest of the service period of duration $(T - t)$. This reliability evaluation is based on the fact that the structural element has survived (no failure) all loads that occurred during time t. Thus the resistance r satisfies the condition

$$r > \max \{f_1,...,f_{N(t)}\}. \tag{15.9}$$

The conditional failure probability becomes

$$1 - P(r > \max \{f_{N(t)+1},...,F_{N(T)}\} \mid r > \max \{f_1,...f_{N(t)}\}) =$$

$$= 1 - \frac{P(\max\{f_1,...f_{N(T)}\} < r)}{P(\max\{f_1,...f_{N(t)}\} < r)} = 1 - \frac{\Phi[\beta(T)]}{\Phi[\beta(t)]}. \tag{15.10}$$

what gives the updated reliability index corresponding to the remaining service life $(T\!-\!t)$:

$$\beta*(T-t) = \Phi^{-1}\left[\frac{\Phi[\beta(T)]}{\Phi[\beta(t)]}\right] \tag{15.11}$$

The updated reliability index $\beta*(T - t)$ is plotted in Fig.15.3 as a function of $t\,/\,T$ for $cT = 1, 10, 100$. It is seen that the updating effect is modest unless $t\,/\,T$ is high.

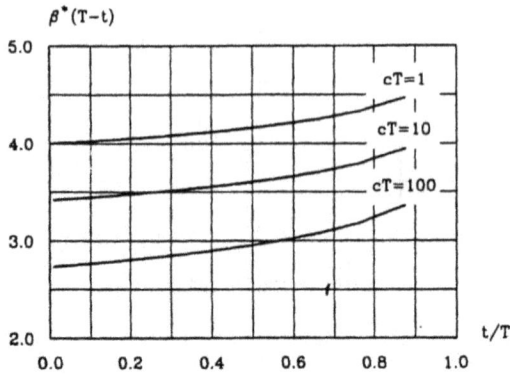

$\beta^*(T-t)$

Fig.15.3 Reliability index in Fig.15.2 updated on the basis of the expression
(15.9)

It is considerably easier to give simple examples when the problem is about updating the exceedance probability for serviceability limit state. A typical example is the use of a sample of measured deformations for given loads. These measurement results can be compared with the results from theoretical models that contain input variables for rigidities of the structural parts and the supports. Another typical example is the use of a measurement of the lowest natural frequency of a mast, for example. The calculated natural frequency depends on input variables that describe the distribution of the masses and rigidities.

Sec.15.3. Decision Criteria

A structure is usually designed in such a way that it satisfies safety requirements of a regulatory body [15.2]. The requirements are related to some formal calculation model and they are based on the information available at the time when the code was enacted. The codified requirements can essentially be reduced to an equivalent requirement that a generalized reliability index $\beta_{a\,priori}$ with respect to a given limit state is not to be less than an expedient reliability index β^{ex} [15.3], where $\beta^{ex} = \Phi^{-1}[1 - P_f^{ex}(T)]$, P_f^{ex} is the expedient value of the probability of failure and $\Phi(x) = \dfrac{1}{\sqrt{2\pi}} \int\limits_0^x \exp(-0.5\xi^2)dx$. The subscript $\beta_{a\,priori}$ indicates that the reliability index corresponds to database available at a pre-design stage of structure development, and thus before any information pertinent to the real operation of the structure hal been available. Data acquisition

250

after inauguration of the structure makes it possible to calculate an up-dated reliability index $\beta_{a\,posteriori}$. Though the code requirement

$$\beta_{apriori} \geq \beta^{ex} \tag{15.12}$$

has a rational background consistent with the pre-existing theoretical cri-teria, more preferable and better acceptable in the public opinion would be a code requirements like [15.3].

$$\beta_{aposteriori} \geq \beta^{ex} \tag{15.13}$$

It will be discussed below that (15.13) as a criterion for not making further actions with respect to the reliability of an existing structure can in fact be based on theoretical considerations. However (15.13) is not a necessary condition for sufficient level of reliability for an existing struc-ture.In order to show the validity of these two statements, possible ap-proaches can be considered that may lead to changes of the existing struc-ture as the consequences of these approaches. In principle there are three basic approaches:
a) no changes to the structure to operate as designed,
b) strengthen the structure and / or change its use,
c) demolish the structure and replace it with a new one.
The b)-type approach may embrace several alternative directions cor-responding to alternative structural strengthening designs and / or service profile. The same holds for the c) – type approach since there may be a choice between different demolition methods and also many alternative designs of the new structure
Let the capital investment in a strengthening system of a given type be given as a function e (β) of the reliability index β for the strengthened structure. This function can be written as

$$e(\beta) = k + h\,(\beta) \quad \text{for } \beta > \beta_{a\,posteriori}$$
$$e(\beta) = 0 \quad \text{for } \beta = \beta_{a\,posteriori} \tag{15.14}$$

where k is a constant= construction cost of the strengthening while g (β) is a increasing function of β defined by $\beta > \beta_{a\,posteriori}$ and starting at the value zero (Fig.15.4). The cost function $e(\beta)$ is not defined for $\beta < \beta_{a\,posteriori}$ because any structural strengthening system has the purpose of increasing the reliabili-ty level beyond the value of $\beta_{a\,posteriori}$. The total expected cost is then

$$k + h\,(\beta) + (d + c_{ny})\Phi\,(-\beta) \tag{15.15}$$

where $\Phi(-\beta)$ is the theoretical failure probability as a function of β while c_{ny} is the optimal investment in a new structure and d is the direct cost of losses due to a failure event. In no strengthening is made, (15.15) is replaced by

$$(d + c_{ny})\Phi\,(-\beta_{a\text{ posteriori}}) \tag{15.16}$$

Comparing (15.15) and (15.16) leads to the following decision rule:
If there is a solution $\beta_0 > \beta_{a\,posteriori}$ to the equation

$$g'\,(\beta) - (d + c_{ny})\varphi\,(\beta) = 0 \tag{15.17}$$

such that

$$k + g(\beta_0) < (d + c_{ny})[\Phi(-\,\beta_{a\text{ posteriori}}) - \Phi(-\beta_0)] \tag{15.18}$$

then a strengthening of the structure should be made. A less expensive alternative design may be considered as well.

If there is no such solution to (15.17) then the considered strengthening design can be dismissed. This does not exclude executing an alternative strengthening design.

If $\beta_{a\,posteriori}$ is considerably smaller than β^{ex} the best decision can be found in choosing a third type approach, that is, demolition and replacement by a new structure.

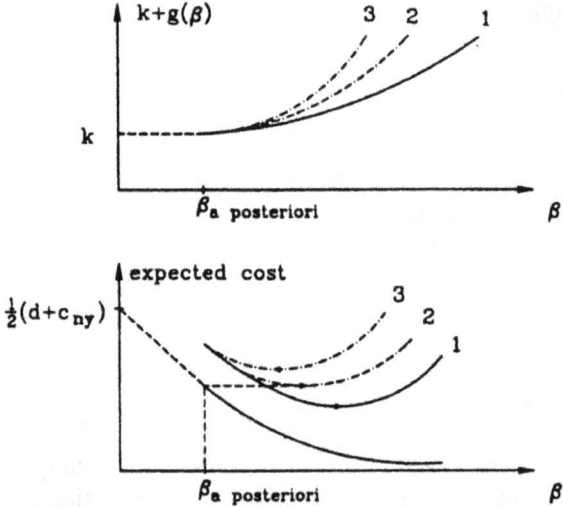

Fig.15.4. Expected cost curve of strengthening.
Upper plot - cost of construction (k) and strengthening (g); Lower plot –
expected cost = loss due to failure (d) + cost of strengthening (c_{ny}).
Curves of the variants listed above: 1 – strengthening; 3 – no strengthening
of the type considered; 2 – shows indifference – another design of strengthening, or demolition, or a new structure, or no changes.
(The curves are schematic)

Let $c_{demolition}$ be the optimal cost for a demolition and let β^{ex} be the optimal reliability index for the new structure with corresponding optimal capital investment c_{ny}. Then it can be concluded that a third type approach is not optimal if the criterion

$$c_{\text{demolution}} + c_{ny} + (d + c_{ny})\Phi(-\beta^{\text{ex}}) > (d + c_{ny})\Phi(-\beta_{\text{a posteriori}}) \quad (15.19)$$

is satisfied. The right side of (15.19) is expected cost of no change to the existing structure. Usually the inequality $c_{demolition} << c_{ny}$ is valid implying that $c_{demolition}$ can be disregarded in the future. Then (15.19) can be rewritten as the criterion

$$\Phi(-\beta_{\text{a posteriori}}) < \Phi(-\beta^{\text{ex}}) + \frac{1}{1 + d / c_{ny}} . \quad (15.20)$$

Unless the right side defines a suitably small probability, the society will hardly accept this criterion. Thus it should be required not only that d is high but also that

$$(d / c_{ny}) >> 1 \quad (15.21)$$

be consistent with the consideration in [15.4] concerning the value of d. It had been disputed that c_{ny} / d should range from 10^{-4} at $\beta^{ex} = 4$ to 10^{-6} at $\beta^{ex} = 5$ which is by an order of magnitude greater than the first term $\Phi(-\beta^{ex})$ on the right side of (15.20). Considering this with an acceptable accuracy, (15.20) can be reduced to the criterion

$$\beta_{\text{a posteriori}} > \Phi^{-1}(c_{ny} / d). \quad (15.22)$$

If this criterion is satisfied neither demolition and a complete renewal should be made. It should be noted that ignoring $c_{demolition}$ and $\Phi(-\beta^{ex})$ in condition (15.19) to (15.22) means that there is no need to change the structure. If (15.22) is not satisfied the choice is either strengthening of the existing structure or total renewal. If (15.22) is satisfied the choice is either no changing the structure or strengthening it.

Finally it will be useful to show that for the practical decision the criterion (15.13) provides a sufficiently accurate condition for doing nothing after updating the reliability. Let k_{min} be smallest value of k for the strengthening systems that are relevant in the given practical situation. Then it is clear that none of these strengthening systems satisfies criterion (15.18) if

$$k_{\text{min}} \geq (d + c_{ny})\Phi(-\beta_{\text{a posteriori}}). \quad (15.23)$$

Thus this condition is a criterion for doing nothing. Also if

$$k_{\text{min}} \geq (d + c_{ny})\Phi(-\beta^{\text{ex}}) \quad (15.24)$$

validity of the inequality (15.13)

$$\beta_{a\ posteriori} \geq \beta^{ex} \qquad (15.25)$$

is another criterion for doing nothing since (15.24) and (15.25) together have the same meaning as (15.23). Since, as previously noted, the right side of (15.24) after division by c_{ny} is of the order of 10^{-1}, the inequality (15.24) is likely to be satisfied in many practical cases. Therefore it can be concluded that the simple canonical solution of the criterion (15.25) in real will often be consistent with making an optimal decision.

A discussion that includes the occurrence probability of errors is given in [15.6]. Criteria (15.22) and (15.25) are unchanged. The sole modification of some importance for the choice of actions is in inequality (15.18). In this inequality the term $\Phi(-\beta)$ in square brackets is the difference between $p_0 = P$ (failure and error for the unchanged structure) and $p_1 = P$ (failure and error for the strengthened structure), [15.3]. The term, $(p_0 - p_1)$, can be expected to be negative because p_0 is an "a posteriori" probability corresponding to the fact that the structure has survived until the present. This fact makes it reasonable to assume that at least some of the possible gross errors that could be made during design, construction and operation of the existing structure actually have not been made and therefore can be disregarded as possibilities. However, there could still exist hidden errors in the existing structure and in addition the errors made during design and construction of the strengthening. This affect is towards increasing the occurrence probability errors with serious implications on the strengthened structure.

References

Ch.10

10.1. Samolinov N.A. (2002), Using non-destructive methods of control of strength of structures in the definition of a residual resource of buildings and structures // Earthquake Engineering. Safety of Structures. №3.(Использование неразрушающих методов контроля прочности конструкций при определении остаточного ресурса зданий и сооружений).

10.2. Rumshinskiy L.Z. (1971), Mathematical processing of the experimental results, M. (Математическая обработка результатов эксперимента).

10.3. Satyanov V.G., Pilipenko PG., Frantsuzov V. A., Satyanov S.V. (2003), Expert examination of industrial safety of industrial buildings and structures ,VISMA, M. (Экспертиза промышленной безопасности производственных зданий и сооружений).

10.4. Satyanov V.G., Pilipenko PG., Frantsuzov V. A., Satyanov S.V., KotelnikovV.S., Method of determination of residual resource of industrial flue and ventilation pipes, Labor Safety in Industry, no.12. (Способ определения остаточного ресурса промышленных дымовых и вентиляционных труб)

10.5. Shmatkov S.B. (2007), Method of calculation of residual service life of building structures, Technical supervision, №5, M.(Способ расчёта остаточного ресурса строительных конструкций).

10.6. Bolotin V.V. (1982), Methods theory of probability and theory of reliability in design of the structures. M. . . (Методы теории вероятностей и теории надёжности в расчётах сооружений).

10.7. Wentzel E.S. (1969), Theory of probabilities, M. (Теория вероятностей)

10..8. Samolinov N.A. (1983), Determination of the stability of the loop develops into account the random nature of the source parameter, Collector, Objects of civil defense. Protective structures, Series 29.73, vol. 2 (56). (Определение устойчивости контура выработки с учётом случайного характера исходных параметров).

10.9. Melchakov A.P. (2006), Calculation and evaluation of accident risk and safe service life of building objects, Publishing house of the SUSU, Chelyabinsk. (Расчёт и оценка риска аварии и безопасного ресурса строительных объектов).

10.10. Popov N.N., Zabegaev A.V. (1989), Design and calculation of reinforced concrete and stone structures, M., 11. SNiP 2.03.01-84*. Concrete and reinforced concrete constructions. - M. (Проектирование и расчёт железобетонных и каменных конструкций).

10.12. Recommendations for definition of terms of service structures prefabricated residential buildings. (1983), APS. M. (Рекомендации по определению сроков службы конструкций полносборных жилых зданий).

10.13. RTM 1652-10-91. (1991), A guide to engineering operation, maintenance and repair of industrial buildings and structures. ICO «NEFTECOM», M. (Руководство по инженерной эксплуатации, содержанию и ремонту производственных зданий и сооружений).

10.14. Position about carrying out of scheduled preventive repair of residential and public buildings (1964), Gosstroy of the USSR, M. (Положение о проведении планово-предупредительного ремонта жилых и общественных зданий).

10.15. Duzinkevich M.S., Lysov D.A., Heiner E.P. (2010), On the strength reserve of load-bearing structures of residential buildings in the first period of industrial house-building, Industrial and civil construction, №12, pp.29-31. M. (О запасе прочности несущих конструкций жилых зданий первого периода индустриального домостроения)...

Ch.11

11.1. General principles on reliability for structures (01.06.1998): ISO 2394:1998(E), Geneva: International Organization for Standardization, 82 p.

11.2. Eurocode-2 (2001), Design of concrete structures. Part 1.General rules and Rules for building: EN 1992-1, (Final Draft),Brussels: European Committee for Standardization, 230 p.

11.3. Basis of structural design. Assessment of existing structures: ISO 13822:2009, International Organization for Standardization, 35 p.

11.4. Eurocode (2001), Basis of structural design: EN 1990, Brussels: European Committee for Standardization, 87 p.

11.5. CONTECVET (2004): A validated user's manual for assessing the residual service life of concrete structures, GEOCISA, 205 p.

11.6. Strength Evaluation of Existing Concrete Buildings (1997): ACI 437R-91 – ACI committee, 437 p.

11.7. ACI Committee 318 (2002), Building Code Requirements for Structural Concrete (ACI 318-02) and Commentary (ACI 318 R-02), American Concrete Institute, Farmington Hills, Mi, p. 444.

11.8. Faber M., Sorensen J.(2002), Reliability Based Code Calibration. Paper for JCSS. Aalborg University, p. 1-17.

11.9. Ditlevesen O.,Madsen H. (2005), Structural Reliability Methods. Technical University of Denmark, 345 p.

11.10. Holicky M., Markova J.(2002), Calibration of Reliability Elements for a Column. JCSS Workshop on Reliability Based Code Calibration, Clokner Institute, STU in Prague, p.1-13.

11.11. Probabilistic Model Code (12th Draft) (2000), Part 1 – Basis of Design: - Joint Committee of Structural Safety – JCSS-OSTL/DIA/VROU, 57 p.

11.12. Sorensen J.(2000), Calibration of Partial Safety Factors in Danish Structural Codes. JCSS Workshop on Reliability Based Code Calibration. University of Alborg, p.1-9.

11.13. Tour V.V., Petcold T.M., Malycho V.V., Markovsky D.M.(2007), Multilevel system of assessing the reliability of reinforced concrete structures of operated buildings and structures. The construction science and technology, 4, p. 4- 19. (Многоуровневая система оценки надежности железобетонных конструкций эксплуатируемых зданий и сооружений).

11.14. Vasiliev B.F., Rozenblyum A.YA. (1974), Reinforced concrete columns of single-storey industrial buildings (calculation and design), M., Stroyizdat Publ. Haus, c.258.(Железобетонные колонны одноэтажных производственных зданий (расчет и конструирование).

11.15. Building Regulations, BR 13-102-2003, The rules of examination of bearing building constructions, M., State Committee of the RF for construction and housing and communal complex (Gosstroy of Russia), 2004, p.27.(Правила обследования несущих строительных конструкций).

11.16. Kolotilkin B.M.(1965), Durability of residential buildings, M., Stroyizdat,Publ. House. (Долговечность жилых зданий).

11.17. RogonskiyV.A. (1969), Mathematical method of evaluation of durability and reliability of elements of buildings, ACADEMYof SCIENCES of the USSR. «Economics and mathematical methods», volume V, issue 1. (Математический метод оценки долговечности и надежности элементов зданий).
11.18. Position about carrying out of scheduled preventive repair of residential and public buildings. (1965), Gosstroy of the USSR, M.,Stroyizdat Publ.House. (Положение о проведении планово-предупредительного ремонта жилых и общественных зданий).

11.19. Position about carrying out of scheduled preventive repair of industrial buildings and structures.(19740, Gosstroy of the USSR, M., Stroyizdat Publ.House. (Положение о проведении планово-предупредительного ремонта производственных зданий и сооружений).

11.20. Kazachek V.G., Lazovskiy N.A.(1997), Actual problems of improvement of operational reliability of buildings and structures, General reports of the International conference «Engineering problems of the modern concrete and reinforced concrete», Minsk. (Актуальные проблемы повышения эксплуатационной надежности зданий и сооружений).

11.21. Kazachek V.G. (20020,Buildings and structures: to prevent an accident during construction and operation, Engineer-consultant in construction, №19.(Здания и сооружения: не допустить аварии при строительстве и эксплуатации).

11.22. Regulations,VSN 58-88 (R).(1990), The regulation on the organisation and carrying out of restoration, repair and maintenance of buildings, objects of municipal and socio-cultural purposes. Design code. State Archit. Committee, M.(Положение об организации и проведении реконструции, ремонта и технического обслуживания зданий, объектов коммунального и социально-культурного назначения).

11.23. Regulations, SNB 1-04.01-04.(2004), Buildings and structures. The basic requirements for the technical condition and maintenance of building constructions and engineering systems, the evaluation of their suitability for use, Ministry of Archit. & Constr. of the Republic of Belarus, Minsk. (Здания и сооружения. Основные требования к техническому состоянию и обслуживания строительных конструкций и инженерных систем, оценке их пригодности к эксплуатации).

11.24. Regulations, TKP 45-1.04-14-2005 (2006), Technical maintenance of residential and public buildings and constructions. Order of carrying out, Ministry of Archit. & Constr. of the Republic of Belarus, Minsk. (Техническая эксплуатация жилых и общественных зданий и сооружений. Порядок проведения).

11. 25. Regulations,TKP 45-1.04-78-2000 (2007), Technical operation of industrial buildings and structures. Order of carrying out. Ministry of Archit. & Constr. of the Republic of Belarus, Minsk. (Техническая эксплуатация производственных зданий и сооружений. Порядок проведения).

11.26 Regulations,SNB 1.04.02-02 (2002), Repairs, reconstruction and restoration of residential and public buildings and constructions., Ministry of Archit. & Constr. of the Republic of Belarus, Minsk. (Ремонт, реконструкция и реставрация жилых и общественных зданий и сооружений).

11.27. Regulations,TKP 45-1.02-104-2008 (2008), Design documentation for the repair, modernization and reconstruction of residential and public buildings and constructions. The procedure of development and approval, Ministry of Archit. & Constr. of the Republic of Belarus, Minsk. (Проектная документация на ремонт, модернизацию и реконструкцию жилых и общественных зданий и сооружений. Порядок разработка и согласования).

11.28. Regulations,TKP45-1.04-37-2008 (2008), Examination of building designs of buildings and constructions. Rules of conduct., Ministry of Archit. & Constr. of the Republic of Belarus, Minsk. (Обследование строительных конструкций зданий и сооружений. Правила проведения).

11.29. Regulations, TKP TN 1990-2009 (2010), Basic principles of design of structures. Ministry of Archit. & Constr. of the Republic of Belarus, Minsk. (Основы проектирования конструкций).

11. 30. Unified codes of depreciation deductions for complete restoration of fixed assets of the national economy of the USSR, Union of Ministry, USSR from 22.10.1990г. № 1072. (Единые нормы амортизационных отчислений на полное восстановление основных фондов народного хозяйства СССР).

11.31. Recommendations to ensure the reliability and durability of reinforced concrete structures of industrial buildings and buildings with their reconstruction and restoration(1990), Kharkov "Promstroyproekt" M., Stroyizdat Publ.House. (Рекомендации по обеспечению надежности и долговечности железобетонных конструкций промышленных зданий и сооружений при их реконструкции и восстановлении).

11.32.Regulations, SP 13-102-2003(2004), The rules of examination of bearing building constructions of buildings and structures, M. (Правила обследования несущих строительных конструкций зданий и сооружений).

11.33 ACI 365. 1R-00.(2000), Service-life Prediction — State of the-Art.Report. ACI committee 365.

11.34Oehme P. (1990),Scheden an Sthahltrawerken Siaristische Schadensanalyse unter Deachtung juristischer Aspekte – Berlin Bauinfrmation, 40 p.

11.35 ISO 15686-1: 2000 (E). Buildings and constructed assets-service life planning (Parts 1-9). 2009.

11.36 Monitoring and Safety Evaluation of existing Concrete Structures. State-of-the-art-Report/ Fib Task Group 5.1, 2002.

11.37. Kazachek V.G.(2010), Foreign experience of standardization of methods of estimation of durability of concrete in the existing construction. Technical regulation, standardization and certification in construction, № 2. (Зарубежный опыт нормирования методов оценки прочности бетона в существующих конструкциях).

Ch.12

12.1. Standard (GOST 27751-88),1988, Reliability of building structures and foundations, M., Publishing house of the standards, 10p. (ГОСТ 27751-88 Надежность строительных конструкций и оснований).

12.2. Code (SNiP II-23-81*) Steel structures, 1991, State Build .Committee of the USSR, M.,96p. (СНиП II-23-81* Стальные конструкции).

12.3. Standard (STO 36554501-014-2008) 2008, Reliability of building structures and foundations. The main provisions, M., SIC Construction, 12p. (СТО 36554501-014-2008 Надежность строительных конструкций и оснований. Основные положения).

12.4. Recommendations (MDS 20-2 .2008), Temporary recommendations on the safety of long-span structures from the progressive collapse in case of accidental actions, SIC Construction, M.,16p. (Временные рекомендации по обеспечению безопасности большепролетных сооружений от лавинообразного (прогрессирующего) обрушения при аварийных воздействиях).

12.5 Rzanitsyn A.R. (1978), The theory of the analysis of building structures on the reliability, M., Stroyizdat Publ. House, 239 p. (Теория расчета строительных конструкций на надежность).

12.6. Raiser V.D. (1998), The theory of reliability in building design, ACV, Publ. House, M., 304 p. (Теория надежности в строительном проектировании).

12.7. Perelmuter A.V.(2007), Selected problems of reliability and safety of building structures. - M., ASV Publ. House, 256p. (Избранные проблемы надежности и безопасности строительных конструкций).

12.8. Perelmuter A.V. (2004), The progressive collapse and methodology of designing of structures , J.Seismic construction. The safety of structures, № 6. p38-41. (Прогрессирующее обрушение и методология проектирования конструкций).

12.9. Shpete G. The reliability of bearing building constructions. - M., Stroyizdat Publ. House, 288 p. (Надежность несущих строительных конструкций).

12.10. Eremeev, P.G. (2009), Protection of long-span structures from the progressive collapse. "The spatial design of buildings and structures",M. , №12, p.209-213. (Защита большепролетных сооружений от лавинообразного (прогрессирующего) обрушения).

12.11. Nazarov YU.P., Gorodetsky A.S., Simbirkin V.N. (2009), To the problem of robustness of building structures on accidental actions, J.Struct. Mech. and Analysis of Struct., №4.,p.5-9. (К проблеме обеспечения живучести строительных конструкций при аварийных воздействиях).

12.12. Almazov V.O.(2008), Problems of progressive collapse of the building objects, Business glory of Russia, p.74-77. (Проблемы прогрессирующего разрушения строительных обьектов).

12.13. Poznyak E.V.(2009), Safety assessment of deformed steel beams on the digital photo image, J.Struct. Mech. and Analysis of Struct., №2, p.67-71. (Оценка безопасности деформируемых стальных балок по цифровому фотоизображению).

12.14. Kirsanov N.M. (1990), Hanging coating of production buildings, M., Stroyizdat Publ. House, 128 p.(Висячие покрытия производственных зданий).

12.15. Streletskyi N.N. (1953), Lattice combined system of bridges, M., Stroyizdat Publ. House, 219 p. (Решетчатые комбинированные системы мостов).

12.16. Shimanovskiy V.N, Sokolov A.A.(1975), Analysis of the hanging structures over the elastic limit, Kiev, Budivelnik Publ. House, 104p. (Расчет висячих конструкций за пределом упругости).

12.17. Ilyushin A.A., Lenskiy V.S. (1959), Resistance of materials, M., Phizmatgiz Publ. House, 371 p. (Сопротивление материалов).

12.18. Recommendations for the choice of types and analysis of the strength of steel ropes, used in the metal structures (1991), Melnikov Design-science Steel-structural Central Institute, M., 32 p.(Рекомендации по выбору типов и расчету прочности стальных канатов, применяемых в строительных металлических конструкциях).

12.19. Dyadkin S.N., Nikolaev V.A., Marikov B.D. (1999), Choice of ropes as the rigging for a road bridge over the Ob river in the district of Surgut, J.

Transport construction, №10, 21-27p. (Выбор канатов в качестве вант для автодорожного моста через Обь в районе Сургута).

12.20. Recommendations on analysis of the steel structures for strength by the criteria of the limited plastic deformations (1985), Melnikov Design-science Steel-structural Central Institute, M. (Рекомендации по расчету стальных конструкций на прочность по критериям ограниченных пластических деформаций).

12.21. Sventikov A.A. (1996), Non-linear analysis of the hanging rod structures, Spatial design of buildings and structures: (research, analysis, design and application): Collected Works, № 8 / Association «Spatial structure»; Belgorod state technological Academy of building materials, M., Belgorod, p.72-82. (Нелинейный расчет висячих стержневых конструкций).

12.22. Sventikov A.A. (2008), Non-linear analysis of spatial hanging rod coatings with the increased stiffness , «Construction and architecture», Scientific Herald of the Voronezh Architectural and Construction University, Issue # 4 (9), Voronezh, p. 27-37.(Нелинейный анализ пространственных висячих стрежневых покрытий повышенной жесткости).

12.23. Sventikov A.A. (2008), Assessment of the reliability of the hanging rod structures in the analysis with the account of geometrical and physical nonlinearity // Computational Civil and Structural Engineering, Volume 4, Issue 2, p.108-109. (Оценка надежности висячих стержневых конструкций при расчете с учетом геометрической и физической нелинейности).

Ch.13

13.1. Risk analysis and safety problems (2007), In 4 parts, part 3. Applied problems of analysis of risks of critically important objects, under ed. Frolov K.V., M., Knowledge. Publ. House, 816 p. (Анализ риска и проблем безопасности. В 4-х частях , ч.3. Прикладные вопросы анализа рисков критически важных объектов).

13.2. Ryzhkin I.I. (2006), Risks of construction and assembly, M., Ankil Publ. House, 248p. (Риски строительства и монтажа).

13.3. Klyuev V.V., Sosnin F.R., Filinov V. N., etc., under ed. KlyuevV.V.(1995), Non-destructive testing and diagnostics M., Machinery Publ. House, 488 p. (Неразрушающий контроль и диагностика: Справочник).

13.4. Grady Butch. Object-oriented analysis and design with examples of applications in C++. Second edition. Rational Santa Clara, California, translation from English, under ed. Romanovsky I., Andreev F.

13.5. Troitsky-Markov I.E., Budadin O.N, Mikhailov S.A., Potapov A.I. (2005), Scientific-methodological principles of energy saving and energy audits. In 3 vol. Vol.1. Scientific-methodological principles of energy audit and energy management, M., Science Publ.House. (Научно-методические принципы энергосбережения и энергоаудита).

13.6. Grunin I.YU., MR. Budko V.B.(2009), Scientific-methodological principles of visual and measuring control in the building examination: textbook under ed. Troitsky-Markov T.E.,. M., VEMO, 166 p. (Научно-методические принципы визуально-измерительного контроля в строительной экспертизе).

13.7. Afanasjev A.A., Matveev E.P. (2008), Reconstruction of residential buildings. Part I. Technology of restoration of the operational reliability of residential buildings, M. (Реконструкция жилых зданий. Часть I. Технологии восстановления эксплуатационной надежности жилых зданий).

Ch.14

14.1. he manual for the design of residential buildings(1989), Vol. 3.Design of residential buildings (to SNiP 2.08.01-85), M., Stroyizdat Publ. House, 304 p.

14.2. Shapiro, G.A., Senderov B.V., Fraint M.YA. (1976), Assessment of the quality of products and installation of large-panel buildings on the results of the strength of full-scale tests, M., Stroyizdat Publ. House, 97p.

14.3. Bayburin A.KH., Golovnev S.G., (2006), The quality and safety of construction technologies, Chelyabinsk, S. -Ural State Univer. Publ. House, 453 p.

14.4. Standard (SNiP 3.03.01-87) (1988), Carrying and protecting structures, M, State Build. Committee of the USSR, 192 p.

14.5. Standard (SNiP 52-01-2003) (2003), Concrete and reinforced concrete structures. The general provisions.

14.6. Standard (GOST R 51901.1-2002)(2002), The management of risk, Analysis of risk for technological systems.

14.7.Standard (GOST 27.310-95) (1995), The reliability in the techniques. The general provisions.

14.8. Senderov B.V.(1991), The accidents of residential buildings, M., Stroyizdat Publ. House, 216 p.

14.9. Recommendations for the prevention of progressive collapse of large-panel buildings (1999), M., 35 p.

Ch.15

15.1. JCSS Publications (2001), Probabilistic Assessment of Existing Structures, 176pp.

15.2. ST 2394 (1998), General Principles on Reliability for Structures, ISO TK98.

15.3. Raizer V.D. (2009), Reliability of Structures.Analysis and Applications, USA, Backbone Publ.Co.,140p. Moscow, 302p., (in Russian).

15.4. Ditlevsen O., Madsen H.O. (1996), Structural Reliability Methods, John Wiley & Sons Ltd., England, 372pp.

15.5. Bendat J.S. and Piersol A.G. (1971), Random Data Analysis and Measurement Procedures, Wiley Interscience, New York.

15.6. Ditlevsen O., Arnbjerg-Nielsen T. (1989), Decision Rules in Re-evaluation of Existing Structures, Proceedings of DABI Symposium on Re-evaluation of Concrete Structures. (Eds.S.Rostam and M.W.Braestrup). Danish Concrete Institute, 239-248pp.

15.7. Rackwitz R., Schrupp K. (1985) Quality Control, Proof Testing and Structural Reliability, J. Structural Safety, 2,239-244pp.

15.8. Madsen H.O. (1987), Model Updating in Reliability Theory, "Reliability and Risk Analysis in Civil Engineering", Proceedings of ICASP 5, Lind N.C. (Ed.), Institute for risk research, University of Waterloo, Vol. 1, 564-577pp.

Table of Contents

Förlag: BoD – Books on Demand, Stockholm, Sverige
Tryck: BoD – Books on Demand, Norderstedt, Tyskland

Signed for printing 22.01.2016. Format 60×90 1/16.
Offset paper. Times type. Offset printing.
Conventional 16,75 printed sheets.

ASV Constraction, Sweden,
Mårdvägen 16 131 50 Saltsjö-Duvnäs

www.ingramcontent.com/pod-product-compliance
Lightning Source LLC
Chambersburg PA
CBHW060241220326
41598CB00027B/4002